BIODIVERSITY: THE UK STEERING GROUP REPORT

VOLUME 1: MEETING THE RIO CHALLENGE

LONDON: HMSO

Printed in Great Britain on Recycled paper.

Dear Secretary of State,

I have had the privilege of chairing the Biodiversity Steering Group and overseeing the preparation of this report.

The signature of the Biodiversity Convention at Rio by the Prime Minister and over 150 world leaders three years ago showed a remarkable commitment to do something to stop the loss of plants and animals and their habitats which were - and are - disappearing at an alarming rate.

The Convention recognised that every country had its part to play in preserving the richness of life. The United Kingdom's response was the Biodiversity Action Plan, published last year, which took stock of the UK's biodiversity and identified a number of ways of doing more to protect it.

The voluntary sector, for its part, produced a comprehensive plan of its own, Biodiversity Challenge, which you welcomed at its launch in January this year.

The Government's Biodiversity Action Plan proposed the setting up of the Biodiversity Steering Group to prepare costed action plans for plants, animals and habitats. The Group is unusual in that it brings together people from a very wide variety of interests including academics, the nature conservation agencies, the collections, business, farming and land management, the voluntary conservation bodies, and local and central Government.

The Group, and its Secretariat, the Sub-Group chairmen and their staff have gone about their work with determination, enthusiasm and energy and I take this opportunity to thank them warmly for their efforts. It was a huge task. For many it has been almost a full time job. There have been differences of view, not so much about the core of our work but about how far we should bring in observations and conclusions which, while important, were incidental to our task. We have erred on the inclusive.

The Report contains in particular:
i   action plans for over 100 species and 14 habitats. We recommend that work on a further 286 species and 24 habitats be completed in the next two to three years;
ii proposals for a United Kingdom biodiversity database;
iii recommendations for raising public awareness of biodiversity;
iv proposals for action at the local level, including guidance on preparing local biodiversity action plans.

We have costed our recommendations, though costing in this area is an inexact science and the figures should therefore be regarded as best estimates only. We have made no assumptions about the availability of public money: it is, of course, for the Government to decide on the pace and priority of this work in relation to other public expenditure priorities. However, it is clear that successful implementation of our plans will depend crucially on all sectors, private, public and voluntary, playing an active part.

How does all this fit into the Government's sustainable development objectives? There is no simple answer, but the health of our biodiversity provides a litmus test for sustainable development. This report offers a way of meeting our Rio commitments and I strongly commend the report to you and your Ministerial colleagues.

Yours Sincerely,
John Plowman.

JOHN PLOWMAN
Chairman

CONTENTS

SUMMARY

INTRODUCTION

1 The Prime Minister signed the Biodiversity Convention at the Earth Summit in Rio de Janeiro in June 1992. At the launch of *Biodiversity:The UK Action Plan* in January 1994, he announced that a Biodiversity Steering Group would be established, with representatives drawn from key sectors and chaired by the Department of the Environment, which would oversee the following tasks:-

- developing costed targets for key species and habitats;
- suggesting ways of improving the accessibility and co-ordination of information on biodiversity;
- recommending ways of increasing public awareness and involvement in conserving biodiversity;
- recommending ways of ensuring that commitments in the Plan were properly monitored and carried out; and
- publishing findings before the end of 1995.

2 Members of the Group were selected on a personal basis to represent central and local Government; the nature conservation agencies; the collections; business, farming and land management; academic bodies and voluntary conservation organisations. It was recognised from the outset that the Group would be advisory in nature, and it is for the Government to respond to the proposals contained in this report.

IMPORTANCE OF BIODIVERSITY

3 There are many reasons why we should conserve biodiversity. We should hand on to the next generation an environment no less rich than the one we ourselves inherited; species which have evolved over many thousands of years may be lost very quickly and cannot be recreated; natural processes help to protect our planet, and in maintaining the productivity of our crops, we depend on a reservoir of wild relatives, and a pool of genetic material that we can go back to, in order to reinforce our selection.

4 We have lost over 100 species in the UK this century, including 7% of our dragonflies, 5% of our butterflies and more than 2% of our fish and mammals.

5 In discharging our obligations under the Biodiversity Convention, the UK Action Plan set as an overall goal:-

"To conserve and enhance biological diversity within the UK, and to contribute to the conservation of global biodiversity through all appropriate mechanisms".

THEMES AND ISSUES

6 We acknowledge that policies are now more sensitive to the needs of biodiversity, but more needs to be done to protect and enhance what we have. To improve air quality, we need to reduce pollutant emissions focusing on photochemical pollution and particulates. The use of water for domestic, agricultural and industrial processes can affect biodiversity in a variety of ways. Systems for producing, transmitting and using energy have a direct or indirect impact on biodiversity, particularly those associated with electricity generation and transport. Agricultural and fisheries practices, minerals and aggregates extraction, urbanisation and an expanding transport network also impact upon biodiversity.

7 In developing individual species and habitat action plans, a number of common issues emerged which, if tackled, would in our view make a major contribution to sustaining and enhancing biodiversity. These include:-

- better advice for land managers on species and site management;
- improved site protection;
- improved grassland management;
- improved or maintained water quality or quantity;
- increased public awareness;
- improved management schemes for woodland;
- the review of agro-chemical use on farmland;
- sensitive development planning and control;
- measures to reduce emission of pollutants;
- further use and refinement of land management schemes; and
- encouragement of inter agency co-operation to conserve habitats below the low water mark.

8 Our analysis underlines the case for continued and vigorous implementation of existing measures, and further development of the many policies which are beginning to incorporate biodiversity objectives. However, in some areas a change of direction may be required, and the targets may not be achievable without a further reform of the CAP involving the redirection of some expenditure to environmental objectives.

TARGETS AND ACTION PLANS

9 In selecting species for which action plans have been prepared, we have used the following criteria:-

- their numbers or range have declined substantially in recent years; or
- they are endemic; or
- they are under a high degree of international threat; or
- they are covered by relevant Conventions, Directives or legislation.

10 Individual action plans, which contain targets for maintaining or increasing the populations and, where appropriate, the distribution of species, have been prepared for 116 of our most threatened and endangered species. We also propose that action plans be prepared for another 286 species within three years, with the great majority within two years.

11 We have prepared action plans for 14 key habitats:-
- for which the UK has international obligations; or
- which are at risk, such as those with a high rate of decline especially over the last 20 years, or which are rare; or
- which may be functionally critical; or
- which are important for key species.

We propose that action plans be prepared for a further 24 habitats within three years, with the great majority within two years.

12 The action plans show the current status of the species or habitat; the main factors which have caused loss or decline; a brief note of what conservation action is currently under way; targets for maintaining or increasing populations and range (species) or size (habitats), and a list of actions which are needed to support the targets.

13 Although the Government, and its agencies, will have ultimate responsibility for delivering the plans, we propose a partnership approach, and the appointment of "champions", who would act as a facilitator and generally stimulate action in taking forward the plan. These could be a statutory body, companies or land managers, or a conservation organisation.

14 We also include a number of habitat statements, to help inform national and local policy and action. These cover the whole land surface of the UK, and the surrounding sea to the edge of the continental shelf in the Atlantic Ocean.

15 If the UK action plan is to be implemented successfully, it requires some means of ensuring that the actions needed at the national level are undertaken in an integrated manner, and that national targets are translated into effective action at the local level.

16 Local Biodiversity Action Plans are seen as a means by which such actions can be achieved. Local Plans should include targets which reflect the values of local people, and which are based on the range of local conditions, and thereby cater for local distinctiveness. They also provide a mechanism for meeting national targets. We believe there is a need for clear guidance on the production of Local Biodiversity Action Plans so that local initiatives are compatible in approach and content. We have therefore drawn up a set of guidelines, and have discussed these with the Local Authority Associations and Local Government Management Board. Agreement in principle has been reached that this approach will be taken forward through the Local Agenda 21 process, and that a series of pilot projects be undertaken to develop Local Biodiversity Action Plans in accordance with the guidance described in Annex C.

17 We recommend a three-pronged approach to improve the quality and accessibility of data and biological recording. This is:-
- making the maximum use of existing data;
- developing a nationally based biodiversity database using a staged approach; and
- developing a locally based biodiversity information system through the establishment of local consortium funding.

18 The sharing and re-use of data should bring economies to national and local bodies. Moving towards a United Kingdom Biodiversity Database (UKBD) will require a staged approach, establishing a co-operative network with a small management centre and investing in products and standards which would encourage data owning organisations to improve the accessibility of their data. We recommend that management of the network should fall to the JNCC.

19 At the local level, we recommend a consortium approach, whereby local data centres would receive funds for providing a service to a range of bodies (eg local authorities, country agencies, NRA, businesses, wildlife trusts, RSPB regions etc).

20 A crucial early stage in developing the UKBD is the creation of a regularly updated directory of datasets and other key information. Although the UK is fortunate in holding a large amount of data on biodiversity (estimated at over 60 million species records), much of this data is very fragmented, and we need to establish who holds what data where. We also need to encourage standards so that information and data are much more compatible.

PUBLIC AWARENESS AND INVOLVEMENT

21 Although the word biodiversity is perhaps a cumbersome and unfamiliar term, it does have some distinct advantages. The conservation of biodiversity is now widely recognised as a new imperative stemming from the Rio Convention. The word biodiversity has been taken by many to mean not simply the variety of life forms on earth, but also the urgent need to ensure their survival. On balance, we consider that the advantages gained from the new connotations associated with biodiversity outweigh the disadvantages of its unfamiliarity.

22 We also recognise that current perceptions of biodiversity, and levels of understanding and interest, vary considerably between different sectors. As a generalisation, the extent to which biodiversity is seen as relevant depends very much on how closely sectors are linked with the natural environment. Biodiversity is seen as highly relevant in sectors where individuals and organisations, such as land managers, have direct contact with the environment. Messages and proposals for action are most likely to be received sympathetically if they come from

leading and respected figures from the sector concerned. We therefore recommend that:-
- key messages be used to raise awareness of biodiversity in each sector;
- "champions" be identified who can act as lead players in each sector;
- "champions" illustrate the importance of biodiversity conservation to that sector by using relevant examples of good practice.

23 The report contains a number of proposals for Government stimulated action; local action; action by key sectors, and education in its broadest sense.

COSTS

24 We have prepared indicative costs for each Species Plan for the years 1997, 2000 and 2010. The costs are for the UK as a whole, and are additional to existing financial commitments. They take account of savings where calculable. We have made no assumptions about where the costs might lie, but note that on past experience half the costs of species action plans have been met from outside the public sector.

25 15 action points have been used to cost each plan ranging from surveys, population studies, special habitat management and habitat creation, wardening of sites, monitoring and advice to land managers. Total estimated costs per annum (of which half are expected to fall to the Government) are £3.8m 1997; £2.9m 2000 and £2.4m 2010.

26 The estimated costs for each habitat action plan cover public expenditure only. These include managing public sector land; the cost of land management scheme payments to private landowners; revenue from land management; and land purchase costs. Total estimated costs per annum are £12.9m 1997; £24.5m 2000, and £37.2m 2010.

27 To put these costs in context, planned 1995/96 public expenditure on agri-environment schemes for the UK is about £100m. This is about 3% of the total UK agricultural support payments of over £3 billion. These are direct payments to farmers plus market support through export refunds etc. In the same year, planned expenditure for English Nature is £41m; Scottish Natural Heritage £40m and the Countryside Council for Wales £17m. Other relevant expenditure includes the National Rivers Authority and Forestry Commission expenditure. Private sector funding is also important, and is thought to run at over £100m annually.

28 The estimated additional expenditure for improving data and information at both the national and local levels is £0.88m 1996/7, £1.1m 1997/8 and £1.3m 1998/9.

29 The work in developing Local Biodiversity Action Plans will form an integral feature of the Local Agenda 21 process. Similarly, the forward programme to increase public awareness and involvement may be taken forward mainly through existing resources, although there is a need to establish a UK Biodiversity Secretariat to co-ordinate some of the work involved. Much can be done through a better focus of current effort.

IMPLEMENTATION

30 To oversee the implementation of the proposals contained in this report, we recommend the following machinery:-
- A National Focus Group:
  which is needed to oversee the preparation of outstanding plans; the implementation of individual species and habitat action plans; to provide advice on good practice; to take forward work on information, data, and monitoring, and to promote public awareness and involvement through action by key sectors.
- Country Focus Groups:
  which may be needed to implement relevant programmes in each country, since institutions in both the private and public sectors tend to be organised at country level.
- "Champions" for each Action Plan:
  who would be invited to act as a co-ordinator for each plan to stimulate the appropriate action.
- Local Biodiversity Action Plans Advisory Group:
  the UK Steering Group for Local Agenda 21 has agreed to establish a small Advisory Group to recommend standards of good practice in the production and implementation of Local Biodiversity Action Plans, and to promote consistency of approach throughout the UK.
- UK Biodiversity Secretariat:
  a small, but full time unit, needed to co-ordinate follow up work; maintain the link with the international convention; service the proposed National Focus Group, and organise workshops and seminars. The unit should be located in DOE.

31 We attach importance to monitoring key species and habitats. The aim should be to establish a cost effective strategy within five years. The monitoring programme will need to be modulated to take account of European and international obligations and existing survey commitments.

32 We see a need for a regular report on progress in delivering the targets, and the implications which arise from monitoring the species and habitats action plans. We propose that this be undertaken every five years by the National Focus Group, with general progress being picked up by the Department of the Environment's annual White Paper.

SECTION I

CHAPTER 1

THE NEED FOR CHANGE

INTRODUCTION

1.1 In January 1994, the United Kingdom Government published *Biodiversity: The UK Action Plan*. This was in response to the commitment given by the Prime Minister at the Earth Summit in Rio de Janeiro in 1992. The United Kingdom was one of the first countries to produce a biodiversity strategy and action plan in accordance with Article 6a of the Convention on Biological Diversity which requires each Contracting Party:-

"to develop national strategies, plans or programmes for the conservation and sustainable use of biological diversity, or adapt for this purpose existing strategies, plans or programmes which shall reflect, inter alia, the measures set out in this Convention relevant to the Contracting Party concerned".

1.2 One of the strengths of the plan is that, for the first time in a Government paper, it draws together existing instruments and programmes for nature conservation throughout the United Kingdom. It commits the UK Government to the strategic objective of conserving and, where possible, enhancing biological diversity within the UK and contributing to the conservation of global biodiversity through all appropriate mechanisms.

1.3 The plan recognises the need for a range of targets for biodiversity to be published in 1995. To advise the Government, and to assist with this work, a Biodiversity Steering Group was established, with representatives drawn from key sectors and chaired by the Department of the Environment.

1.4 As explained in *Biodiversity: The UK Action Plan*:- 'The Group would be advisory in nature and would not affect the responsibilities of the Government, Government agencies or any of the bodies from which members of the Group were drawn. It would be for the Government ultimately to adopt conclusions arising from the work of the Group, and for the nature conservation agencies to discharge their responsibilities within the statutory and financial frameworks provided for them.'

1.5 The targets and proposals contained in this report reflect the considered views of the Steering Group, and as such are put forward to Government as the basis for a co-ordinated and effective forward programme up to the year 2010.

1.6 This Chapter describes the background to biodiversity - what it is, why it is important, and how it is a key component of sustainable development. Chapter 2 explains the development of national targets for key species and habitats and the need for good guidance with the production of local

targets and local action plans. Chapter 3 describes an approach for improving biological recording and Chapter 4 explains the public's perception of biodiversity and how raising public awareness can best be achieved by working with individual sectors.

1.7 Chapter 5 considers main themes and issues. Chapter 6 brings the proposals for action together into a forward programme which refines, prioritises and spells out in more detail the broad targets contained in *Biodiversity: The UK Action Plan*. The remit and composition of the Steering Group are shown at Annex A; guidance in developing Local Biodiversity Action Plans at Annex C, and individual action plans for key species and habitats are included in Annex G.

Song Thrush

THE BIODIVERSITY CONVENTION

1.8 The Convention on Biological Diversity was an important component of the Earth Summit, and was signed at Rio by over 150 countries including the United Kingdom (and the European Community). The United Kingdom ratified the Convention on 3 June 1994.

1.9 Article 1 of the Convention explains its objectives.

"The objectives of this Convention, to be pursued in accordance with its relevant provisions, are the conservation of biological diversity, the sustainable use of its components and the fair and equitable sharing of the benefits arising out of the utilisation of genetic resources, including by appropriate access to genetic resources and by appropriate transfer of relevant technologies, taking into account all rights over those resources and to technologies, and by appropriate funding."

1.10 The Convention reflects a general concern that human activities are changing and destroying habitats, natural ecosystems and landscapes on an increasing scale. It recognises that biodiversity should be treated as a global resource to be protected and conserved according to principles of ecological, economic and social sustainability. Or put more simply, species which take tens or hundreds of thousands of years to evolve naturally can be lost very quickly and cannot be recreated. Their rate of extinction is increasing at an alarming rate, and action is needed to reverse this trend.

1.11 Reflecting this increasing decline in biodiversity in many parts of the world in recent decades, Article 8(f) of the Convention states that each Contracting Party shall, as far as possible and as appropriate:-

> *"Rehabilitate and restore degraded ecosystems and promote the recovery of threatened species, inter alia, through the development and implementation of plans or other management strategies".*

1.12 This theme is in line with a number of others addressed at the Earth Summit. These included:

- the Rio Declaration, a Statement of Principles which addresses the need to balance the protection of our environment with the need for sustainable development;
- Agenda 21, an action plan for the next century which aims to integrate environmental concerns across a broad range of activities;
- a Statement of Principles for the sustainable management of forests, and
- the Convention on Climate Change which commits all ratifying countries to prepare national programmes to contain greenhouse gas emissions, and to return emissions of carbon dioxide and other greenhouse gases to 1990 levels by the year 2000.

Aspen and Oak

1.13 The United Kingdom published strategies for Sustainable Development, Sustainable Forestry and Climate Change at the same time as it published the Biodiversity Action Plan. Although the protection and enhancement of biodiversity is important in its own right, it should be seen as one of a family of initiatives arising from the Earth Summit. The UK plans and strategies are inter-connected. In many Government departments, responsibility for environmental aspects of policy lies with a designated "Green Minister". A Ministerial Committee on the Environment meets to consider questions of environmental policy Government-wide.

1.14 Although Government is in the lead, conserving and enhancing biodiversity requires action by the whole community. Delivering the biodiversity targets contained in this report will require concerted effort from central and local Government, the nature conservation agencies, business, land managers, non-Government organisations (NGOs) and individual members of the public.

WHAT IS BIODIVERSITY?

1.15 Article 2 of the Biodiversity Convention defines biological diversity as:-

> *"The variability among living organisms from all sources including, inter alia, terrestrial, marine and other aquatic ecosystems and the ecological complexes of which they are part; this includes diversity within species, between species and of ecosystems".*

1.16 Put more simply, biodiversity is the variety of life. It encompasses the whole range of mammals, birds, reptiles, amphibians, fish, insects and other invertebrates, plants, fungi and micro-organisms such as bacteria and viruses.

1.17 No one knows for certain the number of Earth's species. Informed opinion suggests a figure of between 5 and 30 million. Our knowledge of the major groups of living organisms in the UK is very uneven. Microbial organisms, such as bacteria and protozoa, are much less studied than flowering plants, birds and mammals. Invertebrates are generally less well known than vertebrates and only very few species have their population sizes assessed. We do however know broadly how many birds, and increasingly, reptiles, amphibians and mammals we have.

Ladys Slipper Orchid

NUMBERS OF TERRESTRIAL AND FRESHWATER SPECIES IN THE UK COMPARED WITH RECENT GLOBAL ESTIMATES OF DESCRIBED SPECIES IN MAJOR GROUPS

(> = more than)
[Source - Biodiversity: The UK Action Plan]

Group	British species	World species
Bacteria	Unknown	>4,000
Viruses	Unknown	>5,000
Protozoa	>20,000	>40,000
Algae	>20,000	>40,000
Fungi	>15,000	>70,000
Ferns	80	>12,000
Bryophytes	1,000	>14,000
Lichens	1,500	>17,000
Flowering plants	1,500	>250,000
Non-arthropod invertebrates	>4,000	>90,000
Insects	22,500	>1,000,000
Arthropods other than insects	>3,500	>190,000
Freshwater fish	38	>8,500
Amphibians	6	>4,000
Reptiles	6	>6,500
Breeding birds	210	9,881
Wintering birds	180	-
Mammals	48	4,327
TOTAL	c.88,000	c.1,770,000

ESTIMATED PERCENTAGE OF SPECIES FROM DIFFERENT GROUPS OF ORGANISMS THOUGHT TO EXIST AS A PROPORTION OF THE GLOBAL TOTAL

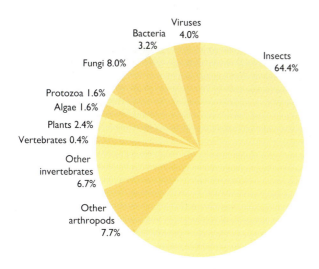

Viruses 4.0%
Bacteria 3.2%
Fungi 8.0%
Protozoa 1.6%
Algae 1.6%
Plants 2.4%
Vertebrates 0.4%
Other invertebrates 6.7%
Other arthropods 7.7%
Insects 64.4%

1.18 However, the concept of biodiversity goes beyond the actual number of species. It includes the variability within a species and the assemblages of plants, animals and micro-organisms which together form ecosystems and natural habitats.

The linkage between species and habitats is important. Nature has an in-built propensity to adjust to change. It is continually evolving new variants and new species. But all this occurs naturally over a length of time. The world is losing species at an accelerated rate as a result of human activity. Evolution cannot keep up and compensate for losses at anything like their current rate, and so biodiversity is declining fast.

GENETIC CONSERVATION

As well as variation between species, there is also variation within species, reflecting their genetic make-up. The genetics of species are shown in a variety of ways - the ability of some populations to live in harsh environments, differences in size or colour, the resistance of some individuals in a population to disease, and so on. Variation at this level is also an important component of the UK's biodiversity, and it needs to be considered from two points of view.

First, conserving the variation of the UK's species. This may be of considerable value in developing locally adapted strains, or for breeding new varieties for commercial reasons, or for conserving local and regional gene pools. The Forestry Commission's work on the native Scots pine *(Pinus sylvestris)* combines these approaches. A group of protected habitats, spanning the natural range of the habitat in the UK, is probably the only way to conserve the genetic variation of species that occur in that habitat type. However, limited research has been focused on the genetic variation of non-commercial species, and there is a risk that groups of sites may not be the best way of conserving the genetic variation of any particular species. Certainly, in any *ex situ* conservation programmes, the ability to incorporate as wide a range of the genetic variation as possible is an important consideration.

Natural selection and random changes in genetic composition in isolated patches of species will drive dynamic changes in their genetic structure resulting in distinctive local populations. Highly mobile pollen clouds may cause mixing of apparently separated plant populations. Animals with large dispersal distances may mix freely.

Second, there is the question of deliberately or accidentally introducing non-local genetic material, either from elsewhere in the UK or from abroad. In the UK there is concern over the importation of wild flower seeds. The focus in semi-natural woodlands is for natural regeneration, and hence the avoidance of introducing non-local genotypes. In the west of Scotland, there is considerable concern about the effects of escapes of farmed salmon *(Salmo salar)* on the wild populations.

The American cord-grass, *Spartina alterniflora*, introduced into Southampton Water, hybridized with the native cord-grass, *Spartina maritima*, resulting in both parent species being swamped by the hybrid, *Spartina x townsendii*, and later by its fertile derivative *Spartina anglica*. Similarly, the establishment of naturalised populations of Sika deer *(Cervus nippon)* in Britain has resulted in hybrids with the red deer *(Cervus elaphus)*; concern has been expressed that in the future all *Cervus* individuals in Great Britain, except for islands where Sika deer are excluded, will contain genes of both species.

While there is an extensive literature on the effects of 'biological invasions' there is only a very limited focus on genetic considerations. The effects of introducing non-local genetic material are poorly understood, but the few case studies available indicate that there could be potential long-term problems for the conservation of the UK's biodiversity.

1.19 Article 10e of the Convention on Biological Diversity says:-

> *'Each Contracting Party shall, as far as possible and as appropriate, encourage co-operation between its Governmental authorities and its private sector in developing methods of sustainable use of biological resources.'*

This concept is an underlying theme of *Sustainable Development: The UK Strategy* (Cm 2426) which seeks to achieve economic development to secure higher standards of living, now and for future generations, while also seeking to protect and enhance the environment now and for our children. In most cases a sensible balance can be sought. But where resources are non-renewable, or the effects may be irreversible, precautionary action will be necessary. In ecological terms this must be before the population of a species falls below a critical level from which it may not recover.

1.20 When considering the levels of use a given area, or a natural resource such as a species, is able to sustain an understanding of its environmental capacity, or carrying capacity, is necessary. Managing the demand for natural resources is one of the fundamental means by which we can move towards a more sustainable way of life.

1.21 We therefore need to work out the environmental consequences of our decisions, and build them into cost benefit analyses and environmental impact assessments. This in turn requires a better understanding of ecosystems and how robust they are in responding to land use change and other impacts and perturbations. While acknowledging that in many cases the interactions are complex, and our knowledge of natural systems is incomplete, it is important that environmental considerations are based, as far as possible, on sound science. It is also important that these environmental considerations should be taken into account as early as possible in the planning and design process so that adverse effects may be avoided or minimised.

Otter

THE IMPORTANCE OF BIODIVERSITY

1.22 No organism lives in isolation from other living things and each has its own way of life which contributes to the balance of nature. The inter-dependence and successful functioning of all these parts is a key contributory factor to the healthiness of the planet as a whole. If we continue to pollute the atmosphere, contaminate land and water, and destroy forests, wetlands and marine environments, we are progressively impoverishing our own environment. We have nowhere else to go.

Limestone Pavement

1.23 Every part of the UK has been affected by human activity, whether directly or indirectly, and most of the landscapes that we now regard as 'traditional' have been influenced or heavily modified by man. This century has seen a formidable increase in the pace and scale of human intervention in the natural world, and the loss of biodiversity has accelerated.

1.24 The underlying reasons why we should conserve biodiversity include a moral obligation, good stewardship, the benefits to society and commercial gain. Taken together they provide a compelling case.

WHY WE SHOULD CONSERVE BIODIVERSITY

Moral and aesthetic
- We should conserve species and habitats because they enrich our lives.
- We should hand on to the next generation an environment no less rich than the one we ourselves inherited.
- The culture of a nation is closely allied to its landscapes and wildlife. Poets, painters, writers and composers have been inspired by the nature around them.
- A culture which encourages respect for wildlife and landscapes is preferable to one that does not.

Stewardship
- Human beings exercise a determinative power over other creatures: with this dominion comes responsibility.
- Species which have evolved over many thousands of years may be lost very quickly and cannot be recreated. Biodiversity cannot be regained overnight.

Benefit to society
- Natural processes help to protect our planet, eg flood plains act as natural release valves for rivers in flood; diversity of vegetation on mud flats and sand dunes reduces coastal erosion; woods and hedges act as wind breaks; upland vegetation is good at binding soil and lessening erosion, and beds of seaweed reduce wave action.
- Wetlands act as natural filters for surface waters and are being used for waste water treatment.
- There is considerable uncertainty about the value of species, for example some plants promise potential cures for different forms of cancer. Since our knowledge is so limited, it makes sense to preserve as many species as possible.

Economic value
- Recent advances in biotechnology have pointed to the potential use of the genetic material contained in plants, animals and micro-organisms for agricultural, forestry, health and environmental purposes.
- To maintain the productivity of our crops, we depend on there being a reservoir of wild relatives, and a pool of genetic material that we can go back to in order to reinforce our selection.
- Much of the countryside in the UK is of great beauty and is a focus for recreation and tourism. For example our National Parks receive well over fifty million visits each year, while research for the British Tourist Authority indicates that overseas visitors are attracted as much by our varied landscapes as by our history and culture.

Medicinal Leech

THE UK ACTION PLAN

1.25 The overall goal in *Biodiversity: The UK Action Plan* is:-

"To conserve and enhance biological diversity within the UK and to contribute to the conservation of global biodiversity through all appropriate mechanisms".

The objectives for conserving biodiversity which underpin this goal are to conserve and, where practicable, to enhance:-
- the overall populations and natural ranges of native species and the quality and range of wildlife habitats and ecosystems;
- internationally important and threatened species, habitats and ecosystems;
- species, habitats and natural and managed ecosystems that are characteristic of local areas; and
- the biodiversity of natural and semi-natural habitats where this has been diminished over recent decades.

1.26 These objectives form the basis of the broad targets contained in *Biodiversity: The UK Action Plan*. These provide a range of programmes and tasks for the Government and its nature conservation agencies, in partnership with others, over the next 20 years. Key components of this programme are:-
- developing costed targets for our most threatened and declining species and habitats;
- improving the accessibility and co-ordination of biological datasets and considering future information management requirements which include the monitoring of agreed targets;
- increasing public awareness and involvement by targeting key sectors; and
- recognising the importance of local biodiversity action plans which complement national action plans - action plans need to be implemented and monitored using both a top down and a bottom up approach.

1.27 The proposals contained in this report go a considerable way in addressing these important issues. But the task is not yet over. The action plan will continue to be dynamic, and there is further work to do in formulating individual plans for the remaining threatened and declining species and habitats identified in Chapter 2. This work should be taken forward within the next three years, in parallel with implementing the other proposals contained in Chapter 6.

SECTION 2

TARGETS FOR KEY SPECIES AND HABITATS

INTRODUCTION

2.1 If we are to conserve and enhance the populations and natural ranges of native species, and the quality and range of wildlife habitats and ecosystems in the UK, we need to develop both national and local targets for our most threatened and declining species and habitats.

2.2 Over 100 species are thought to have become extinct in the UK this century. These include the mouse eared bat; the Norfolk damselfly, the hairy spurge and the summer lady's tresses orchid. Action is required to slow down the loss of our wildlife.

NATIONAL TARGETS

2.3 We have used both species and habitats as starting points. Most species are effectively covered through habitat protection and management, and this is addressed in the habitat action plans. However, some species require their own action plan either because they are found in only a few sites or because, where they are widespread, their special requirements need attention. The quantified targets reflect the judgement of the experts involved, in light of current knowledge. In many cases, they represent a best estimate of an achievable but challenging target, rather than an optimum population or area.

The identification of key species

2.4 The first stage was to identify criteria for selecting key species. The list is based on the best information available. Information on many marine species, lower plants and invertebrates is sparse. It was agreed that species which qualified for one or more of the following categories should be considered:-

- threatened endemic and other globally threatened species;
- species where the UK has more than 25% of the world or appropriate biogeographical population;
- species where numbers or range have declined by more than 25% in the last 25 years;
- in some instances where the species is found in fewer than 15 ten km squares in the UK; and
- species which are listed in the EU Birds or Habitats Directives, the Bern, Bonn or CITES Conventions, or under the Wildlife and Countryside Act 1981 and the Nature Conservation and Amenity Lands (Northern Ireland) Order 1985.

Laurie Campbell

Red squirrel

EXAMPLES OF SPECIES THAT THIS CENTURY HAVE ALMOST CERTAINLY BECOME EXTINCT IN THE UK LARGELY DUE TO HUMAN ACTIVITY (INCLUDES TWO SPECIES SUBSEQUENTLY RE-INTRODUCED)

Species	Suspected main reasons for loss in the wild	Date of last record
Vertebrates		
Burbot (Lota lota)	This fish is assumed to have been lost through river pollution	1972
Mouse-eared bat (Myotis myotis)	Excessive disturbance at, and destruction of, nursery sites	1990
Sea eagle (Haliaeetus albricilla)	Persecution through poisoning, shooting and egg collection. Subsequently re-introduced	1916
Invertebrates		
Large blue butterfly (Maculinea arion)	Lack of grazing and destruction of its grassland habitat. Subsequently re-introduced.	1979
Essex emerald moth (Thetidia smaragdaria)	Coastal defence works leading to a population with a too limited pool for successful reproduction	1991
Viper's bugloss moth (Hadena irregularis)	Loss of Breckland heath to agriculture and development	1979
Blair's wainscot (Sedina buettnei)	The sole site, a coastal marsh, was destroyed by draining and burning	1952
Norfolk damselfly (Coenagrion armatum)	Degradation of the small marshy pools it inhabited through reed, willow or alder growth, or through desiccation	1957
Orange-spotted emerald dragonfly (Oxygastra curtisii)	Pollution from a sewage treatment plant caused its loss from the river where it was most numerous	1951
Exploding bombardier beetle (Brachynus scolopetus)	Lack of management of calcareous grassland	1928
Horned dung beetle (Copris lunaris)	Ploughing-up of pastures on chalky or sandy soils	1955
A click beetle (Melanotus punctolineatus)	The last site was destroyed during golf course construction	1986
Aspen leaf beetle (Chrysomela tremula)	Decline of woodland coppicing	1958
Plants		
Thorow-wax (Bupleurum rotundifolium)	This annual cornfield weed was lost through the improved cleaning of seed corn	1960s
Lamb's succory (Arnoseris minima)	A weed of arable fields, this annual became extinct probably due to agricultural intensification	1970
Hairy spurge (Euphorbia villosa)	Cessation of woodland coppicing at its only site	1924
Summer lady's tresses (Spiranthes aestivalis)	This orchid became extinct as a result of the drainage of the bogs in which it occurred	1959

SPECIES FOR WHICH ACTION PLANS HAVE BEEN PREPARED

Mammals

Arvicola terrestris	Water vole	*Pipistrellus pipistrellus*	Pipistrelle bat
Lepus europaeus	Brown hare	*Rhinolophus ferrumequinum*	Greater horseshoe bat
Lutra lutra	Otter	*Sciurus vulgaris*	Red squirrel
Muscardinus avellanarius	Dormouse	*Phocoena phocoena*	Harbour porpoise
Myotis myotis	Greater mouse-eared bat		

Birds

Acrocephalus paludicola	Aquatic warbler	*Loxia scotica*	Scottish crossbill
Alauda arvensis	Skylark	*Perdix perdix*	Grey partridge
Botaurus stellaris	Bittern	*Tetrao urogallus*	Capercaille
Burhinus oedicnemus	Stone curlew	*Turdus philomelos*	Song thrush
Crex crex	Corncrake		

Reptiles and amphibians

Lacerta agilis	Sand lizard	*Bufo calamita*	Natterjack toad
Triturus cristatus	Great crested newt		

Fish

Alosa alosa	Allis shad	*Coregonus autumnalis*	Pollan
Alosa fallax	Twaite shad	*Coregonus albula*	Vendace

Insects

Argynnis adippe	High brown fritillary	*Formica candida*	Black bog ant
Aphodius niger	A dung beetle	*Formica exsecta*	Narrow-headed ant
Asilus crabroniformis	A robber fly	*Formica pratensis*	Black-backed meadow ant
Bembidion aregentoleum	A ground beetle	*Gryllotalpa gryllotalpa*	Mole cricket
Boloria euphrosyne	Pearl-bordered fritillary	*Hesperia comma*	Silver spotted skipper
Bombus sylvarum	Shrill carder bee	*Idaea ochrata*	Bright wave moth
Callicera spinolae	A hover fly	*Limoniscus violaceus*	Violet click beetle
Carabus intricatus	Blue ground beetle	*Lucanus cervus*	The stag beetle
Cathormiocerus brittanicus	A broad-nosed weevil	*Lycaena dispar*	Large copper butterfly
Chrysotoxum octomaculatum	A hover fly	*Maculinea arion*	Large blue butterfly
Coenagrion mercuriale	Southern damselfly	*Mellicta athalia*	Heath fritillary
Coscinia cribraria	Speckled footman moth	*Oberea oculata*	A longhorn beetle
Cryptocephalus coryli	A leaf beetle	*Panagaeus crux-major*	A ground beetle
Cryptocephalus exiguus	A leaf beetle	*Stenus palposus*	A reed marsh beetle
Eurodryas aurinia	Marsh fritillary	*Tachys edmondsi*	A ground beetle
Eustroma reticulatum	Netted carpet moth		

Other invertebrates

Anisus vorticulus	Snail	*Pisidium tenuilineatum*	A freshwater pea mussel
Austropotamobius pallipes	White-clawed crayfish	*Pseudanodonta complanata*	Depressed river mussel
Catinella arenaria	Sandbowl snail	*Segmentina nitida*	Shining Ram's horn snail
Edwardsia ivelli	Ivell's sea anemone	*Vertigo angustior*	Narrow-mouth whorl snail
Hirudo medicinalis	Medicinal leech	*Vertigo genesii*	Round-mouth whorl snail
Margaritifera margaritifera	Freshwater pearl mussel	*Vertigo geyeri*	A whorl snail
Myxas glutinosa	Glutinous snail	*Vertigo moulinsiana*	Desmoulins' whorl snail
Nematostella vectensis	Starlet sea anemone		

Flowering plants

Alisma gramineum	Ribbon-leaved water plantain	*Fumaria occidentalis*	Western ramping-fumitory
Apium repens	Creeping marshwort	*Gentianella anglica*	Early gentian
Artemisia norvegica	Norwegian mugwort	*Liparis loeselii*	Fen orchid
Cochlearia micacea	Mountain scurvy-grass	*Luronium natans*	Floating water plantain
Coincya wrightii	Lundy cabbage	*Najas marina*	Holly-leaved naiad

Cotoneaster cambricus(integerrimus)	Wild cotoneaster	*Najas flexilis*	Slender naiad
Cypripedium calceolus	Lady's slipper	*Potamogeton rutilus*	Shetland pondweed
Damasonium alisma	Starfruit	*Ranunculus tripartitus*	Three-lobed water-crowfoot
Epipactis youngiana	Young's helleborine	*Rumex rupestris*	Shore dock
Euphrasia spp	Seven eyebrights	*Saxifraga hirculus*	Yellow mountain saxifrage

Ferns

Athyrium flexile	Newman's lady fern	*Trichomanes speciosum*	Killarney fern

Fungi

Battarraea phalloides	A fungus	*Poronia punctata*	Nail fungus
Boletus satanas	A fungus	*Tulostoma niveum*	A fungus

Lichens

Buellia asterella	A lichen	*Pseudocypherellia aurata*	A lichen
Caloplaca luteoalba	A lichen	*Pseudocypherellia norvegica*	A lichen
Collema dichotomum	River jelly lichen	*Schismatomma graphidioides*	A lichen
Gyalecta ulmi	Ulm's gyalecta		

Mosses

Buxbaumia viridis	A moss	*Didymodon glaucus* (*Barbula glauca*)	Glaucous beard moss
Thamnobryum angustifolium	Derbyshire feather-moss	*Ditrichum cornubicum*	Cornish path moss
Weissia multicapsularis	A moss	*Drepanocladus vernicosus*	A moss

Liverworts

Jamesoniella undulifolia	A liverwort	*Marsupella profunda*	A liverwort
Leiocolea rutheana	Norfolk flapwort	*Petalophyllum ralfsii*	A liverwort
Lejeunea mandonii	A liverwort		

Stoneworts

Chara muscosa	Mossy stonewort		

2.5 This approach produced a list of some 1250 species which is shown at Annex F. About 400 species were drawn from this long list. This second list identifies those species which are either globally threatened or are rapidly declining in the UK, ie by more than 50% in the last 25 years.

2.6 From the list of 400 species, a shortlist for early action was drawn up containing 116 species. This third list contains the species for which action plans have been prepared. For more information on the three lists, their contents, and the action plans see Annex F and Annex G.

2.7 The species for which action plans have been prepared have at least one of the following attributes:-
- their numbers or range have declined substantially in recent years; or
- they are endemic; or
- they are under a high degree of international threat; or
- they are covered by relevant Conventions, Directives and legislation.

In addition, we have brought forward a few species with a high popular appeal, such as the otter, which also met one of the above criteria.

2.8 Detailed biological information on each species is not included in the action plans as this can be found elsewhere. The emphasis is on action required and the changes needed to achieve targets. In line with the objectives of *Biodiversity:The UK Action Plan*, targets for species are generally expressed as maintaining status and range or improving these where possible. Numerical status and geographical range both help to cover genetic diversity.

Drepanocladus Vernicosus

2.9 All plans and lists are based on the best available information and will need regular review:-

- the populations of all long list species should be monitored where possible;
- action plans should be prepared for the remaining species on list 2 within three years, with the great majority within two years; and
- work should proceed on refining the lists, with the intention of publishing a further revision by 1997.

Habitat Statements

2.10 *Biodiversity: The UK Action Plan* covers the whole of the UK. A classification of 37 habitat types, to include the whole land surface of the UK, and the surrounding sea to the edge of the continental shelf in the Atlantic Ocean has been developed as a basic framework (some further work is required to fill gaps, eg caves and natural rock exposures). A brief habitat statement has been prepared for each of these to inform national and local policy and action. These are set out in Annex G.

2.11 In selecting this broad classification, two main criteria were used:-

- a workable number of habitat types to ensure the process remained feasible; and
- simplicity - the definitions should be easily understood, unambiguous and recognisable by a broad range of people.

The Identification of Key Habitats

2.12 We decided that habitats qualifying under one or more of the following criteria should be treated as key habitats for costed action plans:-

- habitats for which the UK has international obligations;
- habitats at risk, such as those with a high rate of decline especially over the last 20 years, or which are rare;
- areas, particularly marine areas, which may be functionally critical (essential for organisms inhabiting wider ecosystems) such as sea grass beds (for spawning fish);
- areas important for key species.

2.13 Table 1 shows the broad habitat types in column 1; the key habitats (selected against the criteria explained in the previous paragraph) in column 2, and habitats listed for special attention in the EC Habitats Directive in column 3. Action plans have been prepared for those habitats shown with an asterisk in column 2, and are included in Annex G.

2.14 Costed action plans have been prepared for 14 key habitats, and these are set out in Annex G.

2.15 We propose that:-

- costed habitat action plans be written for the remaining key habitats within three years, with the great majority within two years;
- the list of key habitats be reviewed after three years;
- a list of characteristic species be produced for each key habitat within two years together with indicators of habitat quality; and
- gaps in coverage of representative broad habitat types should be identified and habitat statements prepared by 1997.

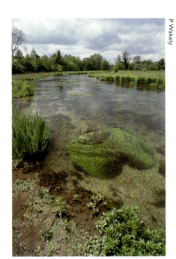

Chalk river

2.16 The Government's Sustainable Development Strategy commits it to publishing a preliminary set of indicators of sustainable development within two years. Publication is expected in early 1996. These will show major indicators relevant to wildlife and biodiversity and to the key impacts on it from man's activities, rather than indicators at the detailed level proposed for key habitats.

Table 1: THE 37 BROAD HABITATS
THE 38 KEY HABITATS
EC HABITATS DIRECTIVE LISTED HABITATS FOR SPECIAL ATTENTION

Broad Habitats	Key habitats (* = costed action plan prepared)	EC Habitats Directive - Annex I types (+ = priority in the Directive)
1. Broadleaved and yew	Upland oakwood* Lowland beech Upland mixed ash woodland Wet woodlands	Old oak woods with *Ilex* and *Blechnum* in the British Isles Beech forests with *Ilex* and *Taxus* rich in epiphytes (Ilici-Fagion) Asperulo-Fagetum beech forests Stellario-Carpinetum oak hornbeam forests Tilio-Acerion ravine forests+ Bog woodland+ Residual alluvial forests (Alnion glutinoso-incanae)+ Old acidophilous oak woods with *Quercus robur* on sandy plains *Taxus baccata* woods+
2. Planted coniferous woodland		
3. Native pine woodland	Native pine wood*	Caledonian forest+
4. Lowland wood pastures and parkland	Lowland wood pastures and parklands	
5. Boundary features	Ancient and/or species - rich hedgerows*	
6. Arable	Cereal field margins*	
7. Improved grassland		
8. Unimproved neutral grassland	Lowland hay meadow Upland hay meadow	Lowland hay meadows (*Alopecurus pratensis, Sanguisorba officinalis*) Mountain hay meadows (British types with *Geranium sylvaticum*)
9. Acid grassland	Lowland dry acid grassland Purple moor grass and rush pastures*	Siliceous alpine and boreal grasslands *Molinia* meadows on chalk and clay (Eu-Molinion)
10. Calcareous grassland	Lowland calcareous grassland Lowland calcareous grassland	Semi-natural dry grasslands and scrubland facies on calcareous substrates (Festuco-Brometalia) (important orchid sites)+ Semi-natural dry grasslands and scrubland facies on calcareous substrates (Festuco-Brometalia) Species-rich *Nardus* grassland, on siliceous substrates in mountain areas (and sub-mountain areas, in continental Europe)+ Alpine calcareous grasslands *Juniperus communis* formations on heaths or calcareous grasslands
11. Lowland heathland	Lowland heathland*	Northern Atlantic wet heaths with *Erica tetralix* Southern Atlantic wet heaths with *Erica ciliaris* and *Erica tetralix*+ Dry heaths (all subtypes) Dry coastal heaths with *Erica vagans* and *Ulex maritimus*+
12. Grazing marsh	Coastal and floodplain grazing marsh*	
13. Fens, carr, marsh, swamp and reedbed	Fens* Reedbeds*	Alkaline fens Calcareous fens with Cladium *mariscus* and carex davalliana+ Petrifying springs with tufa formations (Cratoneurion)+ Alpine pioneer formations of Caricion bioloris atrofuscae+ *Molinia* meadows on chalk and clay (Eu-Molinion) Transition mires and quaking bogs Depressions on peat substrates (Rhynchosporion)
14. Lowland raised bog	Raised bog	Active raised bogs+ Degraded raised bogs (still capable of natural regeneration) Depressions on peat substrates (Rhynchosporion)
15. Standing open water	Mesotrophic standing waters* Eutrophic standing waters Aquifer fed naturally fluctuating water bodies	Natural eutrophic lakes with Magnopotamion or Hydrocharition -type vegetation Hard oligo-mesotrophic waters with benthic vegetation of *chara* formations Oligotrophic waters containing very few minerals of Atlantic sandy plains with amphibious vegetation: *Lobelia, Littorella* and *Isoetes* Oligotrophic waters in medio-European and perialpine areas with amphibious vegetation: *Littorella* or *Isoetes* or annual vegetation on exposed banks Dystrophic lakes

Table 1: continued

Broad Habitats	Key habitats (* = costed action plan prepared)	EC Habitats Directive - Annex I types (+ = priority in the Directive)
16. Rivers and streams	Chalk rivers*	Floating vegetation of *Ranunculus* of plain and submountainous rivers
17. Canals		
18. Montane (alpine and subalpine types)		Alpine and subalpine heaths Sub-Arctic willow scrub Eutrophic tall herbs
19. Upland heathland	Upland heathland	*Juniperus communis* formations on heaths or calcareous grasslands Northern Atlantic wet heaths with *Erica tetralix* Dry heaths (all types)
20. Blanket bog	Blanket bog	Blanket bog (active only)+
21. Maritime cliff and slope	Maritime cliff and slope	Vegetated sea cliffs of the Atlantic and Baltic coasts
22. Shingle above high tide mark	Coastal vegetated shingle structure	Perennial vegetation of stony banks Annual vegetation of drift lines
23. Boulders and rock above high tide		
24. Coastal: strandline		
25. Machair	Machair	Machair
26. Saltmarsh	Coastal: saltmarsh	*Salicornia* and other annuals colonising mud and sand *Spartina* swards (Spartinion) Atlantic salt meadows (Glauco-Puccinellietalia) Mediterranean salt meadows (Juncetalia maritimi) Mediterranean and thermo-Atlantic halophilous scrubs (Arthrocnemetalia fructicosae)
27. Sand dune	Coastal sand dune (including dune grass, dune heath, dune scrub and strandline vegetation)	Embryonic shifting dunes Shifting dunes along the shoreline with *Ammophila arenaria* (white dunes) Fixed dunes with herbaceous vegetation (grey dunes)+ Decalcified fixed dunes with *Empetrum nigrum*+ Eu-Atlantic decalcified fixed dunes (Calluno-Uliceae)+ Dunes with *Salix arenaria* Humid dune slacks Dune juniper thickets (*Juniperus* spp)+
28. Estuaries	Estuaries	Estuaries Mudflats and sandflats not covered by sea water at low tide
29. Saline lagoons	Saline lagoons*	Lagoons+
30. Islands and archipelagos		
31. Inlets and enclosed bays (including sea lochs, rias and voes)	*Ascophyllum nodosum mackii* beds Maerl beds Deep mud Sea grass beds*	Large shallow inlets and bays
32. Open coast	Maerl beds Chalk coasts (littoral and sublittoral)	Sandbanks which are slightly covered by sea water all the time Mudflats and sandflats not covered by sea water at low tide Reefs Submerged or partly submerged sea caves
33. Open sea water column		
34. Shelf break		
35. Offshore seabed		
36. Limestone pavements	Limestone pavements*	Siliceous scree Eutric scree Chasmophytic vegetation on rocky slopes - Calcareous sub-types Chasmophytic vegetation on rocky slopes - Silicicolous sub-types Limestone pavements+
37. Urban		

Note 1 Not all Annex 1 types found in the UK are listed in the table. The reasons for their omission are:-

a UK examples are poor by European standards (see list 1).

b Type is actively controlled (see list 2).

c UK examples are rare and restricted to a single, or a few sites (see list 3), and therefore an action plan/statement seems inappropriate.

d The habitat statements prepared do not adequately embrace certain Annex 1 types (see list 4).

Table 1: continued

List 1 Poor example by European Standards
Marine columns in shallow water made by leaking gases.
Woodland dunes of the Atlantic coast.
Alpine rivers and the herbaceous vegetation along their banks.
Pioneer vegetation and rock surfaces.
Caves not open to the public.
Forests of Ilex aquifolium.

List 2 Invasive species actively controlled in the UK
Dunes with Hippophae rhamnoides.

List 3 UK examples are rare and restricted
Continental salt meadows (Puccinellietatia distantis).
Open grassland with Corynephorus and Agrostis of continental dunes.
Stable Buxus sempervirens formations on calcareous rock slopes (Berberidion p.).
Mediterranean temporary ponds.

List 4 Not covered by existing Habitat Statements
Calaminarian grasslands.
Siliceous scree.
Eutric scree.
Chasmophytic vegetation on rocky slopes - Calcareous sub-types.
Chasmophytic vegetation on rocky slopes - Silicicolous sub-types.

Relationship between species and habitats

2.17 An analysis was undertaken to relate species on the long list to each habitat. It showed that many different habitats contain large numbers of species on the long list, and that actions to conserve UK species will therefore need to be taken in more than just a few key habitats. The analysis also showed that a small number of habitats contain a large number of the species listed, and helped to identify areas of high species diversity.

L Gill

The Noup Noss

WILDLIFE HABITATS WHICH HAVE THE MOST SPECIES OF CONSERVATION CONCERN

An analysis has been carried out for all 37 broad habitats and most key habitats (see Table 1) to see how many species of conservation concern identified in Annex F, they each contain.

The 15 broad habitats which are primary habitats for the greatest number of species of conservation concern are:-

Broad habitat	Number of extant species
Broadleaved woodland	232
Standing open water	136
Natural rock exposure and caves	135
Calcareous grassland	112
Maritime cliff and slope	91
Lowland heathland	82
Running open water	75
Upland heathland	74
Fen and swamp	73
Arable and horticulture	72
Montane	70
Boundaries	65
Open coast	61
Estuaries	54
Sand dune	52

In terms of numbers of threatened species present broadleaved woodland is clearly of high importance, reflecting both its high degree of structure, which provides many ecological niches, and also the fact that it covered most of the country before man started to clear the land for agriculture. It is also the habitat type from which most species have become extinct in the last 100 years, with 46 species being lost, and the one with most globally threatened and rapidly declining species (78 species).

Key habitats other than those shown in the table which support large numbers of species of conservation concern include hedges (47 species), upland oak woodland (47), vegetated sea cliff (45), lowland wet grassland (44), lowland wood pasture and parkland (38), and native pine woodland (37).

No conclusions should be drawn from this analysis about the importance of marine habitats for conservation because pertinent information on species within these habitats is very poor. Also, although some habitats may support only low number of species of conservation concern, nevertheless these species may be of exceptional conservation importance. Other habitats may contain assemblages of plants or animals which are especially species-rich, and so are very important for biodiversity, without containing many special species.

LOCAL TARGETS

2.18 Chapter 7 of *Biodiversity: The UK Action Plan* explained that *"Biodiversity is ultimately lost or conserved at the local level. Government policies create the incentives that facilitate or constrain local action."* If the UK Action Plan is to be implemented successfully, we need to translate national targets into action at the local level.

2.19 Local Biodiversity Action Plans are perhaps the best way forward. One of their main functions is to ensure that national targets for species and habitats are attained in a consistent manner throughout the UK. But there is more to Local Plans than providing a mechanism for meeting national targets. They should be seen as providing the biodiversity element of Local Agenda 21. The setting of targets in the Plan is a crucial part of the agenda for sustainable development, and the extent to which these targets are achieved will be a measure of success. These

Southern Blue Damselfly

targets thus provide a set of indicators for Agenda 21. The development of Local Biodiversity Action Plans should be commended to both public and voluntary sectors as a way of implementing strategic policies for nature conservation within the spirit of Agenda 21.

2.20 One of these objectives is to promote the conservation of species and habitats characteristic of local areas. Local Biodiversity Action Plans provide a focus for local initiatives, and provide an opportunity for local people to express their views on what is important. Local Plans should include targets which reflect the values of local people and which are based on the range of local conditions, and thereby cater for local distinctiveness. However, since they will largely involve land which is in private ownership and management, the approach will require widespread consultation, guidance and involvement, to create the new working partnerships necessary for success.

2.21 A Local Biodiversity Action Plan can act as a catalyst to develop the effective partnerships needed to ensure biodiversity in the long term. These local partnerships may, in turn, raise awareness of the importance of biodiversity, thereby gaining wider public commitment. Widespread "ownership" of a Local Biodiversity Action Plan is regarded as crucial to success in building commitment within the local community.

2.22 Local Biodiversity Action Plans should also identify where it is appropriate to halt recent trends in habitat fragmentation, and create new and attractive landscapes by habitat enhancement and restoration. The data which form an integral part of Local Biodiversity Action Plans need to be compatible with the national biodiversity database, and thus they will play a key role in monitoring progress towards national targets.

2.23 There is a need for clear guidance on the production of Local Biodiversity Action Plans to ensure that local initiatives are compatible in terms of methodology and content. This is so that Local Biodiversity Action Plans can be related easily to the national picture and also to assist local action in this rapidly developing field. In summary:-

the purpose of Local Biodiversity Action Plans is to focus resources to conserve and enhance biodiversity by means of local partnerships, taking account of both national and local priorities.

A general approach and guidance in developing Local Biodiversity Action Plans is explained in Annex C.

CHAPTER 3

INFORMATION AND DATA

INTRODUCTION

3.1 We are fortunate in the United Kingdom that relatively large amounts of data are collected on biodiversity. But much of this is not readily available in a form that assists decisions on the management of species populations or the direction of land-use change. We need to improve the collection, organisation and co-ordination of biological information and data. The opportunity to do this now arises with the need to establish a monitoring programme to measure progress in achieving national and local biodiversity targets. There is also the requirement to monitor progress on the implementation of Directives and Conventions.

3.2 The European dimension is increasingly important with the establishment of the European Environment Agency and its Nature Conservation Topic Centre. The programmes of these organisations are currently being developed. Emphasis is already being placed on the collection and storage of standardised biological data.

3.3 The need to identify important components of biodiversity and for data collection and monitoring under the Biodiversity Convention provides a framework into which requirements from a number of EC Directives and other Conventions fit. The box explains the requirements of Article 7 of the Convention. Table 2 shows the commitment to collect data and information relevant to biodiversity under various EC Directives and the Biodiversity Convention itself. Table 3 shows the current status of monitoring and surveillance for the main groups of species on the long list.

Sand Lizard

ARTICLE 7 OF THE CONVENTION

The starting point for an examination of current practices and future needs is the text of the Convention on Biological Diversity. Article 7 states that each contracting party shall, as far as possible and as appropriate, for the purposes of in situ conservation, ex situ conservation, and the sustainable use of components of biological diversity undertake the following :-

- identify the components of biological diversity important for conservation and sustainable use;
- monitor through sampling and other techniques the components of biological diversity paying particular attention to those requiring urgent conservation measures and those offering the greatest potential for sustainable use;
- identify processes and categories of activities which have, or are likely to have, a significant adverse impact on the conservation and sustainable use of biological diversity and monitor their effects through sampling and other techniques;
- maintain and organise by any mechanism, data derived from identification and monitoring activities relevant to the above measures.

In the identification of biological diversity regard should be given to:-

- ecosystems and habitats:-
  - containing high diversity, large numbers of endemic or threatened species, or wilderness;
  - required by migratory species;
  - of social, economic, cultural or scientific importance;
  - that are representative, unique or associated with key evolutionary or other biological processes;
- species and communities that are:-
  - threatened;
  - wild relatives of domesticated or cultivated species;
  - of medicinal, agricultural or other economic value;
  - of social, scientific or cultural importance;
  - important for research into the conservation and sustainable use of biodiversity, such as indicator species;
- described genomes and genes of social, scientific or economic importance.

Table 2: THE INTERRELATIONSHIP BETWEEN THE REQUIREMENTS TO COLLECT DATA AND INFORMATION RELEVANT TO BIODIVERSITY UNDER VARIOUS EC DIRECTIVES AND INTERNATIONAL CONVENTIONS

Requirement for data and information	EC Birds Directive	EC Habitats Directive	Ramsar Convention	Bonn Convention	Bern Convention	Agricultural Regulations	Water Directives	Forestry Regulations	Fishing Regulations	Biodiversity Convention
Requirement to gather information on:	Wild birds	Habitats and species	Wetlands and species	Migratory species	Threatened habitats	Agricultural practices	Pollution levels	Forestry practices	Yield and population of fish	All components of biodiversity
Maintain and organise data	Yes	Yes	Yes	Yes	Yes	Yes	Yes	Yes	Yes	Yes
Requirement to monitor	Bird population levels	Habitats and species	Wetland and species	Migratory species	Threatened habitats	Various	Water quality	Air pollution effects on forests	Various	All components of biodiversity
Collect information on designated sites for conservation of biological diversity	SPAs	SACs	Wetland sites						Yes	
Data on sustainable use of biodiversity	Yes	Yes	Yes	Yes	Yes	Yes	Yes	Yes	Yes	Yes
Data to quantify threats to biodiversity	Yes	Yes	Yes	Yes	Yes	Yes	Yes	Yes	Yes	Yes

Corncrake

Bittern

Capercaillie

Stone Curlew

28

Table 3: CURRENT MONITORING OF SPECIES ON THE BIODIVERSITY 'LONG LIST'

Group	No. species currently on list	No. species currently with 'biodiversity' action plans	Vagrant species	No. species mainly in wider countryside	No. species mainly on SSSI or reserves	No. species found both in wider countryside and SSSI/reserves	No. species lacking adequate status assessment[1]	No. species with some form of assessment programme in place[2]	No. species which action plans exist or are being considered[3]
Algae (exc. stoneworts)	18	0	0	11	3	4	13	5	0
Fungi	21	4	0	unknown	unknown	unknown	unknown	unknown	6
Lichens	81	7	0	19	37	18	26	44	6
Liverworts	32	5	0	10	11	10	1	30	0
Mosses	79	6	0	27	24	26	20	52	6
Stoneworts	13	1	0	1	7	2	4	9	0
Vascular plants	230	28*	0	70	88	71	9	185	35
Ants	5	3	0	0	1	2	1	4	0*
Bees	19	1	0	5	4	9	0	18	1
Beetles	72	12	0	7	19	42	19	50	3
Butterflies	25	6	0	1	3	21	0	0	25
Caddis Flies	2	0	0	0	0	2	2	0	0
Crickets/Grasshoppers	7	1	0	0	1	5	2	2	3
Dragonflies	7	1	0	0	0	6	2	5	0
Two-winged flies	54	3	0	2	22	28	32	22	0
Mayfly	1	0	0	1	0	0	1	0	0
Millipedes	7	0	0	0	1	6	2	5	0
Molluscs	45	11	0	9	7	29	28	15	2
Moths	122	3	0	27	39	53	11	103	8
Other Invertebrates	40	4	0	17	6	16	23	13	4
Spiders	44	0	0	1	24	17	0	42	2
Stonefly	1	0	0	1	0	0	1	0	0
True Bugs	5	0	0	0	2	3	4	0	1
Wasps	7	0	0	2	3	1	1	5	1
Amphibians	7	2	0	5	2	0	0	5	2
Birds	200	9	3	103	16	77	15	140	43
Fish	25	4	0	10	0	11	1	16	6
Mammals	66	9	17	44	2	2	17	24	7
Reptiles	10	1	3	5	1	1	0	5	5
TOTAL	1,245	116*	23	378	323	454	235	799	166

* The 6 Euphrasia species are covered by one action plan

[1] species poorly known and without current systematic assessment schemes

[2] species for which there are assessment programmes in place

[3] species subject to species action plans, or which are soon to begin. Detailed monitoring assessments are included as part of the plan

3.4 *Biodiversity:The UK Action Plan* explained that the Biodiversity Steering Group would include in its remit overseeing the working group already established following the May 1993 seminar on biodiversity organised by JNCC and DOE. The work of the group would be designed:-

- to improve the accessibility and co-ordination of existing biological datasets;
- to provide common standards for future recording; and
- to examine the feasibility in due course of a single UK Biota database.

Killarney Fern

3.5 In the time available, it has not been possible to validate Table 3 fully. However, the Table does indicate that important groups of species are not covered by the more rigorous forms of surveillance and monitoring included in the last two columns. The proposed Focus Groups we describe in Chapter 6 should be asked as a priority to give consideration to this issue. In doing so, attention needs to be given to the high costs of formal monitoring systems and the lengthy periods before the results can feed into action. There will be the need to prioritise and to consider cost effective approaches including sampling, the identification of groups of indicator species in category one (ie lacking adequate status assessment) which figure significantly on the middle list (see Annex F), and then the possibility of filling major gaps in the long list.

DATA REQUIREMENTS

3.6 The concept of biodiversity, and of national action plans within the framework of the international Convention, has given a focus to a large amount of statutory and voluntary work which has long been continuing and growing. As well as Article 7 of the Convention and the European dimension explained earlier, important aspects are Britain's own Wildlife and Countryside Act, with its schedules of protected species, the EU's Birds and Habitats Directives and many international Conventions -particularly Ramsar, Bonn and Bern. The statutory nature conservation agencies set up under the Environment Protection Act 1990 and the Natural Heritage (Scotland) Act 1991 have a specific duty to protect and enhance nature conservation and are operating major programmes. *Biodiversity:The UK Action Plan*, itself, established broad objectives, set out in the "59 steps", some of which are being carried forward by the work of the Steering Group.

3.7 Data and information are essential if broad aims, specific objectives and precise targets are to be achieved. We need to know where we start from, and what is changing, in order to understand what is causing the change, whether we need or can prevent the change, and to evaluate any remedial action we might take. Programmes, such as the EC Directives and the UK Action Plan, increasingly embody requirements for structured monitoring in which:-

- a baseline is established;
- there is regular and systematic recording to detect change or progress towards specific targets; and
- the reasons for change, particularly undesirable change, are then studied in order to inform action.

Questions of data availability and resources will often make it necessary to compromise, eg by undertaking monitoring of progress towards targets on a less than ideal basis, for example by recording the occurrence of a species in ten kilometre or smaller squares rather than more complete population details.

3.8 In addition, there is a need for broader surveillance which will allow important new trends to be identified, studied and, where appropriate, action taken through operational programmes. An example is the decline of many farmland birds which, while still having large populations, have reduced in numbers by as much as 50% over the last decade.

3.9 The quality, ease of access and relevance of the information available may greatly influence the quality of decision taken. The UK has a large volume of data (estimated at over 60 million species records), but there are important gaps. A growing focus on habitats has exposed a shortage of aggregate data and information on habitats in the UK and EU. A first priority is to improve the accessibility and co-ordination of what we already possess. We recommend mobilising this data applying the principles of information management. Examples of this approach are the IUCN advice on global biodiversity assessment, and the work of the Co-ordinating Commission for Biological Recording (CCBR). The statutory conservation agencies, through the JNCC, and the Biological Records Centre of ITE are the two main focuses of this work, and are already moving in the recommended direction. The local dimension is also crucial.

Sandbowl Snail

RECENT DECLINES IN COMMON FARMLAND BIRD SPECIES IN THE UK

The British Trust for Ornithology's Common Birds Census (CBC) has measured population fluctuations among common species since the early 1960s. It provides an example of how broadly based surveillance can identify important new trends. Fieldwork is carried out by skilled volunteers and covers farmland and woodland habitats using a census method which identifies territorial birds. The CBC provides an estimate of annual change in the size of UK populations of common bird species. In 1994 censuses of the bird populations of 87 farmland plots and 113 plots of other habitats were used to estimate changes.

The declines among farmland birds have been striking as is shown in the table. The declines appear to be driven by the loss of spring-sown cereals and crop rotations, the intensification of grassland management, and the increased use of chemical pesticides; further work is required for a better understanding of these factors and their effects upon birds.

Species	% decline in numbers in 25 years; 1969-1994. Source BTO CBC 'farmland Index'
tree sparrow	-89
grey partridge	-82
corn bunting	-80a
turtle dove	-77
bullfinch	-76
spotted flycatcher	-73
song thrush	-73
lapwing	-62
reed bunting	-61
skylark	-58
linnet	-52

a Note that corn bunting is now so rare that it is found on too few CBC farmland plots to construct a 'farmland' index. Instead CBC plots from all habitats have been used to calculate the decline.

The need for improved information on population trends is reflected in the recent introduction of the Breeding Bird Survey which is jointly supported by BTO, JNCC and RSPB. It is a volunteer-based survey which sets out to increase the coverage of regions, habitats and species over existing schemes, including the CBC, using a formal sampling strategy. Survey methods are simple and efficient, and volunteers record details of both the birds they encounter and the habitats they live in. The new survey provides tremendous potential to identify population declines at a UK or finer level, to provide pointers as to likely causes and either suggest remedial action or identify the need for targeted research.

3.10 The information needed to support the different kinds of decision required by the UK Action Plan is derived from many different data sources held by many different organisations. Examples of areas where information support is required include:-
- selecting species and habitats for action plans;
- establishing targets for them;
- deciding where to direct effort to achieve the targets;
- assessing the effectiveness of action plans; and
- deciding what to do at specific locations.

A SYSTEMS APPROACH

3.11 Information management requires taking a systems approach. The same data will be needed by many people, and much of the information will be re-used many times. The ideal is **to record, check and store once and access many times for many purposes.**

3.12 The steps needed to make our existing information sources more usable and to provide for integrated expansion are:-
- identifying priority datasets for our purposes and indexing the sources;
- establishing standards to ensure that the system works technically, that the information is managed consistently, and that the information is relevant to the needs of customers whether policy makers or those directly concerned with operations;
- capturing accurate data, validating it, storing it securely and keeping a careful record of changes made (in producing managed datasets with such features as version control and "life histories" of the entities recorded);
- generating aggregate information from the original data and ensuring its validity, security and version control;
- avoiding duplication of effort and errors in copying data so that, ideally, each dataset is managed to agreed standards by a single known individual or organisation and made available to others who have a legitimate use for it;

- providing access on appropriate terms in order to avoid duplication and ensure consistency;
- establishing an appropriate charging regime (which could include an absence of charges on a mutual "knock for knock basis" by co-operating organisations);
- making appropriate arrangements for copyright and other aspects of Intellectual Property Rights (IPR);
- conforming to the Environmental Information Regulations.

3.13 At the technical level, the approach recommended below - a disseminated system with potential for electronic networking - is well within the state of the art and does not require massive centralised facilities using experimental designs and equipment.

3.14 At the operational level - service to customers - the risk is of creating something which works but is not useful. Our approach is based on serving policy objectives, including in particular monitoring the progress towards the targets set out in this report and our obligations under UK statute, EC Directives and International Conventions. We start from the use already being made of biological data, and the plans and intentions of many relevant organisations for incremental development. The preferred approach is to establish a basic model of the relevant features (a data model), and to fill in sections of the model as operations require.

Early Gentian

Existing Datasets

3.15 Because of the efforts of dedicated amateurs and, increasingly, statutory organisations, the UK has an exceptionally rich resource of biological data and information. But there are important gaps which it will be slow and expensive to fill, and a significant period is needed before new sets of records can demonstrate changes over time. An important need is to assess the accuracy, reliability and relevance of existing datasets which often reflect individual enthusiasms or past organisational priorities.

3.16 We have therefore tried to address what is practicable in the short term with constrained resources. A key concept is to squeeze every possible ounce of meaning and value from the present bank of data, and from established programmes for recording at planned intervals (time series monitoring).

We therefore have in mind the approach outlined above of working within a framework which is filled in as required. We are setting out on what should be a long term continuous process to monitor UK biodiversity for the purpose of maintaining, and if possible increasing, that biodiversity.

3.17 Two major elements in this process are analysed below:-
- the feasibility of a UK Biodiversity Database (UKBD) based upon a staged approach; and
- giving primary focus to national and local targets, and giving careful consideration to the best use of existing systems. Monitoring a large number of species is potentially expensive and inevitably slow to bear fruit if new systems have to be developed.

The Feasibility of a UK Biodiversity Database (UKBD)

3.18 The UKBD is recommended as the mechanism which will achieve:-

> *"the accessibility and co-ordination of existing biological data, and provide common standards for future recording"* (Biodiversity: The UK Action Plan - paragraph 10.40).

3.19 Any UKBD should involve a network of co-operating organisations who have each agreed to operate to agreed standards, and accept responsibility for maintaining and providing access to the sets of data which they have collected.

3.20 Most problems are concerned with the human, organisational and financial aspects of the system. Not the least of these will be the development of a culture in which individuals and organisations accept the benefits of the maximum degree of data exchange. The concept of custodianship of datasets is critical to the success of a co-operative system and will need careful development under clear Government guidance. However, we start from a dangerous decline in the resourcing of this work at both local and national levels. The local records centre network is not uniformly resourced, and at present centres are closing or barely ticking over. Many of the major national voluntary and statutory organisations are experiencing reduced funding and, in the case of the public sector, are subject to a general freeze on running costs and cuts in those costs in some organisations.

3.21 The heart of a UKBD will be data on habitats, species, protected areas and the status of these entities under international conventions, EC Directives and domestic legislation. The key organisations concerned with this area are the statutory nature conservation agencies and their Joint Committee together with ITE, and in particular the Biological Records Centre, which is part-funded by the JNCC. Organisations such as the British Trust for Ornithology, Butterfly Conservation and other specialist non-Governmental organisations focus on particular species. The Wildlife Trusts and local records centres focus on local and regional needs. The UK is fortunate in still possessing a significant number of individuals who contribute vital data as a result of their enthusiasm for nature.

3.22 The marine area is seen as increasingly important to conservation. The Southampton Oceanography Centre, MAFF, SOAEFD, the Aberdeen and Plymouth Marine Laboratories, ITE and the Marine Conservation Society all hold important data sets, while the JNCC's Marine Nature Conservation Review and Coastal Directories Project are currently improving the baseline data and classification system. NERC's British Oceanographic Data Centre, working on behalf of the EU, have put together a directory of marine environmental data held by UK agencies. This complements a PC based directory, produced by the same group, of some 462 marine environmental datasets held by UK laboratories and readily accessible to both policy makers and scientists.

3.23 Beyond this core area, important datasets are held by various Government departments and statutory organisations. Prominent amongst these are MAFF, particularly through ADAS, and the National Rivers Authority - soon to be part of the Environment Agency.

3.24 As the circle widens, work on biodiversity interfaces with the work carried out particularly by NERC on major processes such as climatic change and atmospheric and marine systems. However, it would be impracticable to try to start by embracing this wider territory. Development of the UKBD should focus on the heartland described above. The wider field should be taken into account, but probably on the basis of specific data transfer provision rather than full compliance with UKBD standards.

The Local Level

3.25 Local centres for data lie at the centre of an effective system. In particular, local data and information are needed to produce Local Biodiversity Action Plans. Local and regional centres both serve the locality and provide information needed for the national picture - the UKBD.

3.26 We gave consideration to funding a fairly uniform network of local records centres on the model which operates successfully in most of the country. However, on reflection we decided to recommend that in each area - to be defined

locally - a consortium should be developed by the leading organisations involved. A more centralised approach would be very costly; would displace existing local resourcing, and would risk reducing local commitment and the relevance of the centres. There is, however, a need for an organisation, or group of organisations, to drive forward work at the local level.

3.27 Where local records centres exist they should be the centre of the network. In other cases the local wildlife trust might be the focus. In all cases the wholehearted co-operation of local authorities is essential, because of the support many of them give to local records centres, and the data they generate or collect, and the use they make of biodiversity data for planning and other purposes. The statutory nature conservation agencies are already providing resources to some centres, but continued local Government funding is critical.

Reed Beds

The National Level

3.28 At the national level we see a network of organisations which co-operate voluntarily. The key concept will be "custodianship". Organisations owning relevant datasets will agree to maintain and develop these sets and to conform with the agreed standards and protocols of the network. There would need to be a focus for the network. If this is to work effectively, it should be provided by one organisation. This should be the JNCC with advisory machinery representing other major players. This is because of JNCC's expertise, its large number of datasets and planned developments, and its long standing involvement with the Biological Records Centre. A small full time headquarters will be required to mastermind the project, to develop standards and to plan and monitor the progress of the UKBD.

The Data Catalogue

3.29 A crucial early stage in developing the UKBD is the creation of a regularly updated directory of datasets and other key information. An important option is to develop this simple text directory into a more technical meta database giving full details of the technical characteristics of the entries. This role should include:-

- assessing the quality of data;
- encouraging improvements, and
- resisting both duplication and redundancy.

Recorder

3.30 An important contribution can be made at both national and local levels by maximising the use of the Recorder software system for recording species and their geographical locations. The system, developed by the statutory nature conservation agencies and the Wildlife Trusts, can be used by individual volunteers or local centres and allows data from appropriately designed surveys to be available in an electronic form which is easy to collate nationally. The package includes aids to standardisation such as a species dictionary. The value of the system to individuals and small organisations would be increased if it could be loaded with appropriate national information, such as species of conservation concern or action plans, and could therefore provide both feedback and context for local work.

Sma' Glen

RECORDER

Recorder is a computer application aimed at people and organisations collecting or collating site based observations of terrestrial and freshwater species. The Recorder project is managed by the Joint Nature Conservation Committee, and the Countryside Council for Wales, the Department of the Environment (Northern Ireland), English Nature, Scottish Natural Heritage and the Wildlife Trusts.

This sample printout illustrates four features of Recorder which would help a locally based organisation meet Biodiversity Action Plan information needs. The printout is from a copy of Recorder used to collate dragonfly data for North East England.

RECORDER	**Demonstration**		**Friday 15 October 1993**
Cordulegaster boltonii (Donovan, 1807)	**Odonata Cordulegasteridae**		

Golden-ringed dragonfly

[1]Large, black and yellow hawker dragonfly. Breeds in acidic streams and, sometimes, lakes and ponds on heaths and moorland, but a very strong flier which can occur far from breeding areas. Common in the west of Britain, especially in south-west England, Wales, Cumbria and north-west Scotland, also more locally in the Midlands and much of the Pennines. Rather scarce in the east and absent from the south-east except for isolated heathland colonies.

Status	**Region**[2]
Local	Northern England
Common	Wales
Common	Scotland
Local	Great Britain

Site[3]	**Grid ref**	**Date**	**Source of record**
Close House	NZ1265	10 July 1981	Ball, Dr S G[4]
Colster Cleugh	NY99	13 July 1973	Long, Dr A G
Shull	NZ0833	02 July 1955	Anon (1955)
Shull	NZ0833	07 June 1947	Anon (1947)
Newton Hall-Pity Me	NZ2745	1961	Heslop-Harrison, J

[1] **Nationally supplied information on species.**
Recorder contains a dictionary of 36.000 terrestrial and freshwater species typically providing information on status, biology and distribution.

[2] **Nationally supplied status and facilities to add local species status.**
National status information provided as part of Recorder can be compared with assessments of local status made by the user. This comparison is one of the steps recommended during Local Biodiversity Action Plan construction.

[3] **Facilities to add site based species records.**
With appropriate data, the package allows the comparison of status of species across suites of sites. This is an important operation when identifying and planning action for locally important species. Around 350 copies of Recorder are in use and capturing a significant proportion of locally generated species data. Users are spread across local records centres, county wildlife trusts, the statutory conservation agencies, the National Trust, several National Parks, individuals and a variety of natural history societies.

[4] **Support for national data collection.**
The application helps users produce highly checked data on the location of species in a consistent electronic format. This has already made the collation of national data sets more efficient for some species groups, and is used by some national species recording schemes.

The Countryside Information System (CIS)

3.31 The Countryside Information System, developed for DOE by ITE to display the results of the Countryside Surveys, is a valuable tool in presenting data which can be set out on a grid of one kilometre squares. Data may include, for example, ecological characteristics, land use, species distributions, altitude, soils, administrative areas and designations. This facility allows broad patterns of these data to be displayed and compared at a national level. The system is designed to provide easy, flexible access to data and a particular feature is the ability to extract information for any configuration of one kilometre squares. CIS provides a standard format for the exchange and dissemination of information, and it contains a catalogue of available datasets.

3.32 Standards are once more the key here. Custodians of information have a range of tools to assist access to their most used datasets eg; CIS, UKDMAP and in the near future the facilities of the emerging Internet. Datasets need to be documented if they are to be used and interpreted easily. With appropriate standards a dataset could be captured in Recorder, collated nationally and subsets published in CIS or UKDMAP. So long as data are geographically referenced (by use of the national grid, latitude and longitude or other information which can be turned into co-ordinates), it can be presented and analysed by the CIS (or a Geographical Information System (GIS)). Particular attention needs to be paid to the requirement of people or organisations with limited competent skills through "user friendly" programs and "front ends" which assist the ordinary user.

High Brown Fritillary

COUNTRYSIDE INFORMATION SYSTEM

The Countryside Information System (CIS) was commissioned by the Department of Environment to provide policy advisers, planners and researchers with an easy and flexible means of accessing a wide range of information on the environment. The CIS can help meet biodiversity commitments in two principal ways:-
- by helping the exchange and linking of information about patterns of biodiversity and environmental impacts.

The CIS is currently being used by a wide range of users ranging from Government departments and local authorities to universities and wildlife non-governmental organisations. In addition to land cover and vegetation data from the Countryside Survey 1990, new information is currently being made available on the system including Ordnance Survey topography and geographical reference data, designated areas and the distribution of farm types in England and Wales. Additional information on birds, flowering plants, invertebrates, soils and climate are expected to be available in 1996.

- by providing a tool for publishing the results of national surveys and monitoring exercises and by promoting the use of common standards of data analysis and presentation.

The CIS is a Windows-based software package which can store, analyse and present maps, tables and graphs for any data that can be summarised for one kilometre squares on the National Grid of Great Britain. It now includes a catalogue which provides detailed information on the system's available data and suppliers.

Standards

3.33 Standards lie at the heart of a UKBD. They fall into three areas:-
- biodiversity standards.
- information standards;
- technology standards;

3.34 Biodiversity standards relate to the content of the databases - species dictionaries, habitat classifications, standardised scales for assessing the status of species and habitats, criteria for identifying species under threat, etc.

3.35 Information standards relate to such matters as definitions of the types of information needed, the structuring of these types for effective management in a database and version control.

3.36 Technology standards relate to the workings of the information technology. For example, communication between modern databases can be achieved by specifying that they all use, or at least can be accessed through, Structured Query Language (SQL). Similarly, there need to be standards for communications on the network. It is likely that these would involve the use of the Internet.

3.37 Work to develop standards is in hand in both the voluntary and statutory sectors. We have begun the major task of mapping this activity and encouraging filling gaps. If appropriate standards are conformed to, data can be recorded, stored and transmitted in a variety of ways extending from paper and post through the exchange of computer prints or floppy disks and electronic networks. The degree of electronic sophistication depends on costs and benefits.

Charging

3.38 An important aspect of the Environmental Information Regulations 1992, (implementing EC Directive 90/313/EEC on the freedom of access to information on the environment), is that public sector holders of relevant data have both an obligation to provide access (with reasonable charges) and to ensure that the data is accurate.

P Wakely

Starfruit

3.39 We consider that the objective should be to provide the most open possible access compatible with these policies and with the need to provide an incentive to data custodians to perform their fairly onerous duties. In some cases, custodians will themselves have strong interests in obtaining data from others and one could think of a "knock for knock" approach. In other cases, perhaps the more commercially orientated public bodies and the private sector, the key datasets are almost by-products of their main activities and they need relatively little reciprocity from other custodians.

3.40 Data providers have had to live with this situation for some time. In the environmental field, ITE, for example, have different rules of access and charging for those who co-operate in exchanges of data, bona-fide researchers and those seeking to use data for commercial purposes. Collaborators or researchers may often be charged only handling costs of data. Access can be controlled through various procedural and electronic devices which, at a cost, provide "gateways" ensuring that rules for access and charging are applied.

Intellectual Property Rights

3.41 The importance of intellectual property in the biodiversity area has recently been emphasised by the Co-ordinating Committee for Biological Recording (CCBR) and others when analysing the Environmental Information Regulations. A crude summary is that recorders of data and organisations collating datasets acquire Intellectual Property Rights which may pass to their estate. These issues have not yet been tested in the courts and various practical steps are being taken to gain clearance for the use of data, but it is necessary to recognise

that this is laborious, time consuming and not without risk of expensive legal judgements.

CONCLUSIONS

3.42 We consider that a UK Biodiversity Database (UKBD) would be an important tool for carrying forward the Biodiversity Action Plan and other commitments under UK and EC legislation and international conventions. At relatively little cost, it would add considerable value to the high volume of existing and planned data.

3.43 The UKBD would be a cost effective way to set standards, improve access and encourage greater use of the existing and planned sources of data and information.

3.44 The UKBD is best thought of as a network of collaborating organisations conforming to an agreed overall design and IT, information management and conservation standards. The work programme could be developed through stages which each involved a product of value in its own right.

3.45 An effective network would require a small management centre, essentially concerned with negotiating standards and participation, which should be based on the JNCC. Contributing organisations would become "custodians". This would involve an undertaking to maintain specified datasets to UKBD standards and to make them available on agreed terms of access and payment.

3.46 The local level is crucial both as a provider of data and a user of that data. It is particularly important to encourage a healthy local sector against a background of decreasing resources and closing local record centres.

3.47 Both national and local custodians may need a financial incentive to participate in the UKBD network. While many custodians in national organisations concerned with biodiversity would expect to benefit, others might find themselves in a one-sided relationship in which they provided more than they used.

3.48 There are intricate questions of charging, rights of access and intellectual property rights which need to be addressed as an early priority.

3.49 We have focused on improving access to existing or planned datasets of importance to biodiversity. We noted, however, that knowledge was limited on important species and habitats, and in particular that programmes of systematic surveillance or formal time series monitoring have limited coverage. Further consideration should be given to this issue, and to any necessary supporting research by the implementation machinery proposed in Chapter 6.

CHAPTER 4

PUBLIC AWARENESS AND INVOLVEMENT

INTRODUCTION

4.1 *Biodiversity: The UK Action Plan* commented that:-

> "Public awareness and appreciation of biodiversity appears to be growing significantly as a result of a wide variety of initiatives. Government policy is seeking to strengthen and extend existing good practice, fill gaps, improve what is ineffective and give the participating bodies a sense of belonging to a national and international movement with well defined aims and a common purpose."
>
> It goes on:-
>
> "biodiversity is ultimately lost or conserved at the local level".

4.2 The Steering Group was asked to consider the preparation and implementation of a campaign to increase public awareness of, and involvement in, conserving UK biodiversity. However, since a number of initiatives were already under way, such as Local Agenda 21, Going for Green, and European Nature Conservation Year 1995, the Steering Group considered that it would not be appropriate to launch another campaign. We did, however, recognise the need for a more focused programme of action to meet specific objectives.

4.3 *Biodiversity: The UK Action Plan* provides a significant opportunity for new approaches to nature conservation to be developed and implemented. The Government's commitment to the conservation of biodiversity, and the need for this to be linked to other initiatives in sustainable development, provide an opportunity to take stock of how biodiversity is regarded within society as a whole and to make recommendations to ensure that the UK Action Plan can be implemented effectively. Public attitudes to biodiversity conservation will be crucial to this process. The Government leadership in promoting greater public awareness of the importance of biodiversity will be crucial to success through bringing on board those people and organisations who, as yet, see the conservation of biodiversity as being of little relevance to them in their daily lives.

PERCEPTIONS OF BIODIVERSITY

4.4 In order to understand current perceptions of biodiversity, and to identify barriers and constraints, we commissioned research to assess the perception of biodiversity in specific sectors. The findings indicate that people's understanding of what biodiversity means falls into three broad categories; species, the natural environment, and sustainability. The most common understanding of biodiversity is "diversity of life", and the idea that species are vital to the maintenance of the Earth's ecology. Biodiversity is also seen to mean the natural environment in a broader sense, including natural habitats. In both cases it seems that implicit in the word "biodiversity" is the need to take action to preserve the widest possible range of species and habitats. Biodiversity is also understood by some to have a deeper and more fundamental meaning linked to the impact of human activities on a global scale. In this sense, biodiversity is regarded as a crucial part of sustainability and is closely linked with Agenda 21.

Lichens

THE ROLE OF KEY SECTORS

4.5 Our research showed that levels of understanding and interest varied considerably between different sectors of society. It is therefore important to gear messages and proposals for involvement and action, in accordance with the particular interest of each sector. A list of key sectors is shown in the attached box.

KEY SECTORS

Action to raise public awareness of biodiversity would best be achieved by working through individual sectors. The key sectors are:-

- land managers;
- business and industry;
- financial institutions;
- the churches and faiths;
- youth and community groups;
- the professions:
  - architecture;
  - science and engineering;
  - planning;
  - legal;
  - landscape design;
  - medical;
- the media;
- all forms of education;
- environmental groups;
- central Government;
- local Government.

4.6 Messages and proposals for action are most likely to be received sympathetically if they come from respected figures in the sector concerned. Champions of biodiversity should be encouraged to provide leadership and inspiration. They should be challenged to supplement the key messages produced in this report with statements of why biodiversity is important to their particular sector.

4.7 The extent to which biodiversity is relevant to various sectors of society varies greatly and appears to depend very much on how closely such sectors are linked with the natural environment. Biodiversity is seen as highly relevant in sectors where individuals and organisations, such as land managers, have direct contact with the environment. Where people have no obvious connection with the land, they do not generally perceive biodiversity to be relevant to them. This is the case with the media, commerce and industry, and youth and community sectors. In some cases, it is seen as a negative factor, as in the case of some businesses in which the most obvious association with biodiversity is felt to be the development restrictions placed on green field sites.

BARRIERS AND CONSTRAINTS

4.8 A number of issues are perceived as constraints or barriers preventing effective action to conserve biodiversity (although some of these refer to the environment, there are direct links with biodiversity):-

- most important of all, the relevance of biodiversity is misunderstood, and regarded as a concept for which a narrow series of actions will suffice. Yet, unless biodiversity is promoted in terms of sustainability, all environmental actions will continue to be essentially palliative;
- biodiversity needs to be seen as a crucial issue in its own right, and as a critical part of sustainability. Significant progress will not be achieved until the issues become sufficiently well defined to provide a basis for lifestyle decisions on the part of individuals;
- the public is often not aware of the connections between day to day behaviour and the consequent effect on the environment. The link between consumption and production has become hidden from view;
- although many companies do make considerable efforts, and incur substantial costs, industry and commerce (in general) are not felt to be predisposed towards environmental issues; with an emphasis on short term profit, environmental features are perceived to be a costly burden;
- while a wide range of organisations produce environmental or conservation material, this information is not always practical enough nor does it always reach those practitioners who need it;
- while a number of environmentally positive activities are occurring across a range of sectors, taken as a whole these appear to be piecemeal and unco-ordinated.

4.9 Public support for biodiversity is most likely to be harnessed if the practice and policies designed to raise public understanding of the term capture the enjoyment people experience from direct contact with the living world; the sense of security people gain from knowing that wildlife is being protected for their children and for future generations to enjoy, and the moral correctness of conserving wildlife for its own sake.

4.10 There is the dichotomy that people want to see nature protected, they want direct contact with it, they want to conserve it for its own sake and for future generations and they want central Government to take a lead, especially in ensuring that industry and commerce comply with environmentally sensitive measures designed to reduce the rate of harmful change to wildlife. But they are unsure about the Government or industry's commitment to deliver effective change.

DEVELOPING A CO-ORDINATED PROGRAMME

4.11 To raise public awareness, a number of specific issues need to be addressed. First there is the matter of the word "biodiversity" itself. Although it is a cumbersome and unfamiliar term, it does have some distinct advantages. The conservation of biodiversity is now widely recognised as a new imperative stemming from the Rio Convention. The word biodiversity has been taken by many to mean not simply the variety of life forms on earth, but also the urgent need to ensure their survival.

4.12 On balance, we consider that the advantages gained from the new connotations associated with biodiversity outweigh the disadvantages of its unfamiliarity.

4.13 Second, we consider it necessary to produce a concise and well argued account of why the conservation of biodiversity is important. This should include both philosophical and the practical issues, together with arguments for the involvement of a wider spectrum of society. Such a statement could then be used to develop a set of key messages aimed at raising awareness. Messages would need to take account of current levels of perception, and be tailored to the many different sectors involved.

4.14 We recommend that:-
- key messages be used to raise awareness of biodiversity in each sector;
- champions be identified who can act as lead players in each sector; and
- they illustrate the importance of biodiversity conservation to that sector by using relevant examples of good practice.

We need to convey the importance, and even the excitement, of biodiversity. Members of the UK Round Table on Sustainable Development may have a role to play here.

4.15 The involvement of local communities including land managers and businesses, will be crucial to the successful implementation of Local Biodiversity Action Plans. There is a need to encourage such activities to ensure the effective implementation of national policies at the local level.

4.16 We therefore propose a co-ordinated programme to raise public awareness and to encourage involvement. Four main categories have been identified where action is required. These are:-
- Government stimulated action;
- local action;
- action by key sectors;
- education in its broadest sense.

The proposals for action which make up this forward programme are explained in Chapter 6, and are set out in Annex D.

4.17 This programme should be implemented **over the next three years,** and should be given high priority by the Government alongside other aspects of the commitments given in Biodiversity: The UK Action Plan. With regard to Education and Training, it is recognised that a longer timescale is required, and we recommend a medium term programme for developing strategies for environmental education covering all parts of the UK over the next 5 to 10 years.

KEY MESSAGES

4.18 Most definitions of biodiversity are couched in scientific language. This does not increase public awareness or understanding of biodiversity. We need to raise public awareness through simple messages written in plain English explaining what biodiversity is, why it is important and emphasising the need to take action to conserve biodiversity.

4.19 Production of an inspirational layman's guide is an important part of this process. Such a statement needs to refer to the great variety of life comprising biodiversity. In effect, it includes all forms of life on earth, but biodiversity goes beyond everyday concepts of nature by including all genetic variation too. A popular statement should emphasise that we have a responsibility for ensuring the future survival of biodiversity, and that we need to take action now. This may be a topic which can be developed as part of the "Going for Green" campaign.

Limestone Pavement

4.20 With these considerations in mind, we have produced several versions of a popular statement which can provide the basis for key messages (see boxes). The three versions are of different levels of complexity, and are intended to meet different needs. A wide range of sectors should be targeted with the object of encouraging greater awareness of the need for conservation of biodiversity and the active involvement of both individuals and organisations.

BIODIVERSITY: THE VARIETY OF LIFE

Biodiversity is all living things, from the tiny garden ant to the giant redwood tree. You will find biodiversity everywhere, in window boxes and wild woods, roadsides and rain forests, snow fields and the sea shore.

But don't take plants and animals for granted. We are part of biodiversity and depend on it for our quality of life. And what we don't save now, our children will pay for later. Biodiversity is the living bank that everyone should invest in. Now it is banking on you.

BIODIVERSITY

Biodiversity represents the amazing richness and variety of plants, birds, animals and insects that exist throughout the world.

No organism exists in isolation from other living things, and each contributes to the balance of nature and the very survival of the planet. Not only the endangered and the exotic, but the everyday, from an antelope to a bog ant, a mahogany tree to a meadow flower.

Human activities are changing and destroying biodiversity on an increasing scale - the way we use farmland and forests, or where we build houses, hotels, marinas and motorways. If we do nothing to prevent this, we will suffer both economic and spiritual loss, and hand on to our successors a planet which is poorer than the one we were privileged to inherit.

Biodiversity provides the raw ingredients for our quality of life - our food and clothing, health and relaxation. What we don't protect now will affect the lives of our children and grandchildren.

So make biodiversity matter and make a world of difference

WHAT IS BIODIVERSITY?

The word gives expression to the amazing richness and variety of different plants, insects, birds and mammals and all living things throughout the world. This is part of our natural heritage, but it is a matter of the gravest concern that this richness is fast declining, mainly because the places in which these plants and animals make their homes are being progressively destroyed by all kinds of human influences.

In the United Kingdom, we have our share of this richness, but here too it is being steadily eroded. Why is it important that we should halt this loss? There are several reasons:-

- all our food comes from plants and animals. We make some use of native wild plants and animals, but by far the greater part is provided by crops and farm animals. All these have originated from wild ancestors by generations of selected breeding. So people have always depended on the diversity of wildlife. Future generations will need new sources of food stuffs, and should be able to turn to the great reserve of living things for this purpose. Also to improve the resistance of our crops and stocks to pests and diseases, we often need their wild relatives to support the genes for resistance which we can breed into the domesticated varieties;
- although we often take them for granted, wild plants and animals are an integral part of our surroundings, providing essential sources of pleasure, interest, knowledge, understanding and relaxation, without which we should be much the poorer. They create the woods, fields, wetlands and moors that we all enjoy. The more we reduce the richness and diversity of living things, the sooner these assets will disappear.

So we must:-
- protect species not only the rare and endangered ones but also the common ones, to prevent them becoming rare;
- protect natural and semi-natural habitats which provide their locations within which biodiversity can be maintained;
- provide areas where people can have access to nature to enjoy contact with the natural world, even in towns and cities.

EDUCATION AND TRAINING

4.21 Education and training influence public awareness and the extent to which the public become involved in action to conserve biodiversity. For a number of reasons, this area has been treated separately from other sectors. This is partly because of the well established processes of formal education, but we are also conscious of the very considerable range of initiatives being pursued in the broad arena of environmental education. We have endeavoured to review the current position as it relates specifically to biodiversity, and to identify a number of issues which merit further action.

4.22 During the past year, environmental education has been considered as a priority topic by the UK Panel on Sustainable Development, and this subject is also in the forefront for consideration by the UK Round Table. Important statements on future policy have been issued by the Department for Education, jointly with the Department of the Environment, and by the Scottish Office.

4.23 In February 1995, the Department for Education and the Department of the Environment organised a joint conference to consider practical support for schools and teachers in providing education about the environment. The conference discussed ways of helping schools to identify and learn from existing good practice, and materials and resources appropriate to the teaching of environmental matters through the National Curriculum. There will be no further change in the National Curriculum for five years so there is no prospect of increased coverage of environmental matters within that timescale. However, as a result of the conference, the Government has decided to give practical help with delivering environmental education in the following three ways:-

- the Council for Environmental Education has been asked to advise Ministers on the practical implications of establishing a Code of Practice for environmental teaching materials and resources, perhaps including a system of peer review;
- a study has been commissioned to examine the feasibility, and the cost and funding implications, of a national index of materials and resources suitable for teaching about environmental matters; and
- the School Curriculum and Assessment Authority, working in partnership with the Council for Environmental Education and the Royal Society for the Protection of Birds has been asked to identify and publish examples of good practice in environmental education.

4.24 These new initiatives will allow opportunities to consider ways in which biodiversity can be dealt with more adequately as part of the National Curriculum. In addition, the Teacher Training Agency is undertaking a fundamental review of in-service training which will take account of the conference report.

4.25 In Scotland, the Secretary of State has recently endorsed the report *Learning for Life* produced in 1993 by the Working Group on Environmental Education. This report sets out a strategy for taking forward environmental education in Scotland over the next ten years. Public consultation following the report supported the proposals which are based on a requirement for environmental learning through life - from cradle to grave. The philosophy embodied in the report provides a holistic approach to environmental education which is widely supported by many practitioners throughout the UK. *Learning for Life* reinforces that environmental education should be delivered wherever and whenever it is relevant to the learner - from the first year of childhood, where the values which will influence the way we live are being developed, to later years when we make choices as parents, workers and consumers, all of which affect the sustainability of our environment.

4.26 The publication of *Learning for Life* has already led to a range of new initiatives in Scotland, some notable features of which are the development in community and informal education, and new initiatives in the work place. A very successful example is the Environmental Community Chest, piloted by Scottish Natural Heritage, which provides a

Balerno School Community Chest

set of resources designed for use by Community Education Services. Another successful initiative is the development of Regional Environmental Education Forums generally based on individual Education Authority Regions. These provide important networks linking into the Agenda 21 process.

Scottish crossbills

4.27 In adopting *Learning for Life* as the strategic base for developing policies which relate to environmental education, the Scottish Office has produced a strategy document which addresses how this approach would be implemented in many walks of life.

4.28 The Council for Environmental Education Biodiversity Working Group strongly endorse the *Learning for Life* model emphasising that biodiversity education is an essential component of broad environmental education comprising awareness, literacy, responsibility and competence. Strategies and plans for environmental education covering all parts of the UK should be produced which would address delivery in the formal, non-formal and informal sectors building on existing mechanisms to produce a more cohesive approach.

4.29 An important element of training is the need for continuing professional development in all sectors which have an influence on biodiversity. There is a need to promote such training through the Professional Institutes in fields such as Architecture, Engineering, Planning, Landscape Design and Chartered Surveying.

4.30 We are aware that there is widespread concern about the adequacy of environmental training for teachers, especially at Primary level. Lack of competence and unfamiliarity with nature lead to a lack of confidence on the part of many primary teachers in dealing with this subject. It is important, therefore, that the current review of in-service training should specifically address biodiversity. In England and Wales, the Grants for Education, Support and Training (GEST) programme continues to support in-service training in LEA maintained schools in basic curriculum subjects. Schools may use School Effectiveness Grants to address their own priorities, set out in their development plans or post-inspection action plans. LEAs may also support designated courses to enhance primary teachers' subject knowledge in all basic curriculum subjects. The mechanism exists, therefore, to provide specific training on biodiversity, and we recognise that biodiversity needs to be given higher priority for the targeting of grants. Increasing awareness of, and competence in, biodiversity should be a criterion for grant aid.

4.31 In England the Department for Education has commended the 1993 Toyne Report on Environmental Education in further and higher education as a basis for colleges and universities to review how the environment is treated within their curricula. It is largely for the institutions themselves to implement the Report's recommendations, but the Department for Education is considering monitoring national progress.

4.32 At a practical level, there is a problem of adequate training in taxonomy which is essential for work on biodiversity. The importance of systematic biology research, and the natural history collections on which this research relies, was recognised by the House of Lords Select Committee on Science and Technology in their report *Systematic Biology Research* (HL Paper 22 of 1991-92). The committee expressed concern over the diminishing level of systematic expertise in the UK and of the decline in training at the school, undergraduate and postgraduate levels. Part of the Government's response to the Report's recommendations was the establishment of the UK Systematics Forum in February 1994 for an initial period of just over two years.

knowledge and experience of biodiversity on an everyday basis. This should lead to a greater sense of responsibility, and direct involvement in environmental issues.

Laurie Campbell

Brown Hare

4.33 The Forum consists of a committee of representatives from major collections-based institutions in the UK which reports to all levels of the systematics community, as well as policy makers and funding bodies. As part of its remit, the Forum aims to promote awareness of the training problems in taxonomy. An important step will involve utilising information compiled in a database of UK systematic expertise and research to provide evidence of the problem and help to identify gaps where skills are needed. Discussions at regional and national meetings have helped raise awareness of the problem and have identified the need to co-ordinate and prioritise at a national level.

4.34 There is a crucial need to re-establish contact between people and nature at first hand - by this we mean people of all ages - as part of *Learning for Life*, in order to encourage direct

SECTION 3

CHAPTER 5

THEMES AND ISSUES

INTRODUCTION

5.1 In recent years, considerably more attention has been paid to environmental issues with consequent benefits for biodiversity. Two examples are agri-environment schemes to promote farming practices that protect and enhance the environment which have been introduced in the UK and throughout the European Community, and the Countryside Stewardship scheme which supports the management and restoration of important habitats, wherever they occur. There are many other examples. However, while policies are now more sensitive to the needs of biodiversity, more needs to be done to protect and enhance what we have.

5.2 The UK has a rich and characteristic flora and fauna for a group of islands of its size. This wealth of plant and animal species is a product of the UK's geographical position, its diverse geology and geomorphology, its soils and the human management of land over many centuries. The biodiversity of the marine environment around the UK, at the juncture of a number of major biogeographic regions, is important in European and Global terms. Within a relatively small area, enormous variations of habitat and species exist, representing a large proportion of the total biodiversity of these islands.

5.3 The number and range of plant and animal species in the UK has been affected by human activities. In the last 50 years, there have been many losses as a result of urbanisation, agricultural and industrial development, an expanding transport network, pollution of air, water and soil, and mineral extraction.

5.4 The Government has a three pronged approach to conservation: designation and protection of specific geographical areas; specific actions targeted towards individual species, and measures aimed at the wider countryside. The key issues are:-

- to conserve and enhance, as far as possible, the wide variety of species of flora and fauna found in the UK, particularly those where the population is of international significance;
- to ensure that the UK's objectives in landscape and wildlife conservation are given their full weight in policies for other sectors, particularly transport planning, industry, forestry, land use, agriculture and coastal protection; and
- to ensure that the level of exploitation of commercially-used species is sustainable, and compatible with their conservation.

5.5 This chapter is concerned with the use of those natural resources which contribute to biodiversity. Our aim should be to ensure that through their wise use, biodiversity is not only protected, but is conserved and enhanced for current and future generations. It should be an aim of policy to minimise further losses of biodiversity due to human activity and, where practical, to increase biodiversity.

GENERAL PRINCIPLES AND ISSUES

Carrying Capacity

5.6 We need to know how much environmental use an area can sustain without damage -its environmental carrying capacity. Levels of exploitation and forms of management need to ensure that use is kept within sustainable limits if biodiversity is to be conserved. Remedial measures need to be agreed and implemented should the limits be exceeded to allow biodiversity to recover.

Precautionary Principle

5.7 To define environmental carrying capacity, we need to understand the vulnerability of populations and ecosystems to land use and other changes. In many cases the interactions are complex, and our knowledge of natural systems is incomplete. In line with the precautionary principle (see glossary) where the available evidence suggests that there is a significant chance of damage to our natural heritage, conservation measures may be appropriate even in the absence of conclusive scientific evidence that the damage will occur. Ensuring that biodiversity is adequately protected can often be achieved at a low cost if protection is planned at the start. Adding measures at a later stage can be prohibitively expensive. Opportunities may also arise to enhance the environment as a condition of the 'development licence' if the biodiversity needs are included at the initial design stage. Some examples are providing coppice for energy; aggregate extraction leaving lakes for wildfowl/ pond life; quarrying leaving cliffs, and imaginative development of road verges.

Lichen Lungwort

Polluter Pays Principle

5.8 We need to consider the impact on the environment of what we do. Those who impose environmental costs on society through environmental damage or the use of important environmental resources, should strive to minimise these costs or compensate for them through environmental enhancement or direct financial payments.

5.9 For the most part we do not put a price on biodiversity. It is treated as a free gift like the air we breathe. So we may fail to place a sufficiently high value on conserving biodiversity. Those processes which consume scarce natural resources, cause pollution or involve the direct loss of wildlife habitat are undercosted. Examples include the use of water resources, the generation and use of energy, and transport systems, particularly roads and the traffic they carry.

Air Quality

5.10 Good air quality is essential for human health and the wellbeing of the environment as a whole. There have been major changes in UK air quality since the 1950s. Widespread coal burning in the home has declined substantially with the move to cleaner fuels such as gas and electricity. Cleaner fuels, less polluting processes and pollution abatement equipment have all contributed to lower emissions from industry. However, motor traffic has increased substantially to the extent that vehicles are now a major source of pollutants in urban areas, but with their effects being felt more widely in the form of acid rain, and through their contribution to climate change. Nitrogen and sulphur dioxides, carbon dioxide and monoxide and other volatile organic compounds can all have impacts on biodiversity both locally and in more widespread ways. The key issues are to reduce pollutant emissions to improve air quality, both locally and for biodiversity generally.

5.11 The areas adversely affected by acid rain and low level ozone can be shown through the use of critical loads. These are estimates of exposure to pollutants below which significantly harmful effects on sensitive elements of the environment are unlikely to occur. For a given pollutant, the critical loads for a particular effect (eg acid damage to trees or eutrophication) can be mapped. If this map is then compared with another showing deposition loads, the areas of exceedance can be seen. Use of this technique has shown that there are currently large parts of the UK where terrestrial and freshwater ecosystems are at risk of acidification, eutrophication or ozone damage as a result of air pollution. Since the pollutants which cause these effects can travel long distances in the atmosphere the solution lies in the substantial reduction of emissions of the pollutants all across Europe. Emissions from power generation, transport, industry and agriculture may all need to be reduced, in some cases by large percentages, before critical load exceedance in the UK and elsewhere are reduced to a more acceptable level.

Water

5.12 The use of water for domestic, agricultural and industrial processes can affect biodiversity in a variety of ways. During prolonged periods of low rainfall, the head waters of some rivers may dry out especially during late summer and autumn. This effect will be compounded by excessive river and groundwater extraction for industrial and domestic supplies and agriculture, with a serious consequence for freshwater and other wetland habitats such as marshes and water meadows already stressed by drought.

5.13 Some plant and animal life in rivers has been adversely affected in the past by certain types of watercourse management and river engineering, for example the straightening of rivers, dredging, and removal of broadleaved trees and shrubs from river banks. A return to greater variety of riverside habitats and vegetation, including stretches with trees and shrubs to provide dappled shade and to contribute organic matter to the river, would enhance aquatic biodiversity.

Mesotrophic Standing Water

5.14 Plant and animal life is also affected by point or diffuse pollution arising from a number of sources, such as effluent from sewage treatment works and industrial processes, leaching from old and unused mines, run-off from agricultural chemicals or farm waste and chemical spills. If groundwater becomes polluted, the contaminants may persist for some time, and springs can be polluted for many months. Problems include low dissolved oxygen levels, high nitrate and pesticide concentrations, heavy metal contamination from mine drainage and fuel spills.

Energy

5.15 Most systems for producing, transmitting and using energy have a direct or indirect impact on biodiversity, particularly those associated with electricity generation and transport. The combustion of fossil fuels to produce energy releases carbon dioxide, the main cause of the greenhouse effect, and sulphur dioxide which contributes to acid deposition. If, as seems highly probable, it is established that climate change is likely to result from greenhouse gas emissions, there will be significant implications for biodiversity. Changes in the distribution and composition of plant communities are predicted.

CLIMATE CHANGE IMPACTS ON BIODIVERSITY

Climate scenario for 2050

Current predictions suggest that by the year 2050 the UK may be on average 2°C warmer, with up to 10% more rainfall. Warmer temperatures are likely to occur in winter and during the night. Increased rainfall is more likely in winter and in the west. Despite more rainfall, higher rates of evapotranspiration may increase soil moisture deficit and drought, particularly in the east. The sea level may rise by 20cm. To imagine the effect of this change, think of the climate of Bristol becoming more like that of Rennes, London more like Paris and Newcastle more like Hamburg. However there is a high degree of uncertainty with this sort of regional scenario and there are many other possible outcomes including the UK being affected by changes in the North Atlantic oceanic circulation which could become unstable thereby reducing the warming effect of the North Atlantic Drift. In the draft reports completed so far, climate change is seen as a relatively unimportant factor over a timescale of 5-20 years (ie 2000-2015).

Direct effects on Ecosystems

The response is likely to be complex, as a result of the combined effects on individual species, competition, local extinction, colonisation, soils and hydrology. Ecosystems will respond to increased temperatures, reduced incidence of frost, reduced moisture availability and the effect of increased carbon dioxide.

Species of montane habitats at the southern limits of their range (eg dotterel) may become extinct in Britain. Conditions for species at their northern limits (southern European species/frost-susceptible species such as the Dartford warbler and chalk hill blue butterfly) may become more favourable. Wetlands in the east and north and upland blanket bogs could become drier, favouring colonisation by grasses and trees.

There will be limited opportunities for plants and animals to colonise new areas because semi-natural habitats are generally fragmented (and as the UK is a group of islands) and because the projected rates of climatic change exceed those for which natural ecosystems are adapted. As conditions become increasingly unfavourable to species in their existing sites (due to competition) they may disappear locally. Some species (eg: some birds, larger mammals and annual weeds) are better adapted for change than others - these species are also the ones which have coped best with past agricultural change and are the least threatened.

Unless action is taken, rising sea level will be associated with increased coastal and river mouth flooding. Mudflats and saltmarsh will be 'squeezed' between existing (and reinforced) sea defences and the rising sea level. However, in some places, increased sea cliff erosion may introduce new sediment to the coastal system and cause localised deposition which may be environmentally beneficial. The Birds in Estuaries research project has shown how the numbers of shorebirds are related to shore width and could be used to help predict the decline in bird populations associated with sea level rise. The RSPB has estimated that 1.2% of wintering waders are 'at risk' from rising sea level and that some nesting waterfowl species (eg avocet) are threatened with extinction.

A research project into The Changes in Key Habitats is assessing the possible effects of climate change (among other factors) on lowland heaths, calcareous grassland, uplands, coasts and watersides in England.

Indirect effects of climate change

The most significant indirect effect will be change in agricultural land use and land management. This may happen both because suitability and profitability for crops and animal production will change in the UK as a result of the local climate, but also because global changes in food production will affect the demand for UK produce. A small amount of agricultural land may be lost due to coastal retreat. The UK may be in a position to increase production (and profitability) to make up for deficits elsewhere. However, much will depend on agricultural policy (such as price support, set-aside and agri-environment schemes), international trade agreements and developments in biotechnology which may reinforce or counter these trends.

Modelling studies as part of the DOE Core Model project suggest that the centre of cereal production may shift towards the north and west, into currently mixed and livestock farming areas. Grassland productivity may improve in currently marginal upland situations supporting more stock and making reclamation of moorland more viable. Soils and slopes will be a limiting factor. There may be opportunities to increase production of crops which are currently marginal in southern England, such as sunflowers and vines. Such changes in agricultural land use will have effects on wildlife through reclamation of previously marginal and semi-natural land, conversion of grassland to arable, increased irrigation, changing patterns of farming such as autumn sowing of cereals and effects on the profitability of intensive farming. The effects of policies for agricultural support and environmental production will, as now, be critical in determining the changes in land use.

Other less significant land use effects may relate to the increased demand for water sports and a revival in the popularity of British coastal resorts in a warmer climate giving rise to increased pressure for recreation and development on wetlands and coastal habitats. Activity related to winter sports in the Scottish Highlands may decrease.

Transport

5.16 A good transport system is essential for a prosperous and growing economy. However, changed verge management, replacement of roadside ditches by buried drains, road construction, the upgrading of existing roads and other transport infrastructures can cause loss of, and damage to, wildlife habitats and wider landscape features. Besides the direct impact of land take, there are can be impacts through increased demand for aggregates, increased pressure for adjacent development, and through the substantial contribution transport makes to greenhouse gas emissions and acid rain.

Farming

5.17 Almost 77% of the UK's land surface is in farming use. Agriculture is therefore a key determinant of biodiversity, and farmers and land managers are vitally important in implementing measures to maintain and enhance biodiversity. Today's countryside has been shaped and maintained largely by farming activities, and even semi-natural areas depend on the continuation of certain traditional forms of agricultural production.

R Key

Lundy Cabbage

5.18 The farmed landscape is an intricate patchwork of semi-natural and highly managed areas supporting a wide range of habitat niches and some important species which depend on them. However, changes in farming, particularly the trend to regional specialisation and more intensive management of crops and pasture have led to the loss of many former farmland habitats from large areas, and reduced the value of much productive land to wildlife, to the concern of many farmers and land managers as well as conservationists.

5.19 There are five main issues:-
- the deterioration of many semi-natural habitats such as chalk grasslands and heather moorland as traditional forms of management become increasingly uneconomic or difficult to continue, and are either abandoned or replaced with more intensive management;
- the loss and fragmentation of semi-natural habitats such as heathland moorland, hay meadows and wetlands, as some of the land is brought into more productive agricultural use, or converted to other non-agricultural uses such as plantations, built development etc;
- the loss of important farmland features, such as hedgerows, ancient trees, copses, ponds, ditches and small wetlands as farms are reorganised to become more economic and take advantage of modern technology;
- the deterioration in the value of productive cropped and grazed land to wildlife as the intensity of production has increased, involving such changes as loss of winter stubble, fashions for crop and grass monocultures, and the switch from hay to silage; and
- damage to food chains and the soil, water and other ecosystems components of farmland caused by atmospheric pollution and pesticides and fertilisers.

Woodlands and Forests

5.20 All woodlands and forests have some value for wildlife, but the remains of "natural" forest cover (the ancient semi-natural woodland - defined as woodland that is known to have existed continuously in England since before 1600 and before 1750 in Scotland) are the most valuable and diverse, and are of special importance because they cannot be replaced. New planting since the First World War has seen the woodland area of the UK expand to about 10% of the land surface. Much of the new woodland area comprises non-native, conifers but the planting of broadleaved species has expanded significantly in the last few years.

5.21 Recently planted woodlands are less diverse, immature ecosystems, although they can add to the biodiversity of a previously unwooded environment, especially land of low wildlife value. New woodlands established close to existing ancient, semi-natural woodland, and woodlands that follow closely the natural processes of succession, and those which develop a spatial and structural pattern which mimics nature, have the greatest potential benefit for wildlife conservation. Single trees and hedgerows trees can also make a contribution to biodiversity.

5.22 There are four main issues:-
- avoiding any further reduction in the area of ancient and semi-natural woodland, which only amounts to 1.4% of the total woodland area and is greatly fragmented;
- the loss of biodiversity through the replacement of habitats of high wildlife value by plantations with less value;
- loss of biodiversity through inappropriate woodland management, or lack of management. There are significant areas of woodland whose value for timber production as well as for wildlife is deteriorating; and
- failure of woodland regeneration due to over-grazing by sheep, deer and cattle. This is of particular concern in some upland areas, where deer and livestock grazing is causing deterioration of woodland, and a reduction in the diversity of woodland flora, but is also a growing problem in the lowlands where deer populations are increasing.

Urban

5.23 About 10% of the land surface of the UK is in urban use. Development is by no means continuous, and within urban areas there is much open land in parks, open spaces, road corridors, private gardens and other areas which contribute to biodiversity, and with appropriate management could enhance it.

5.24 Many urban areas contain relics of natural habitats which have survived development. Parks and private gardens can be important for wildlife, and are the main day to day contact points with wildlife for many people. Given the right conditions, some wildlife can thrive in towns. This can help to raise awareness of the natural world and a concern for its conservation.

Fen Orchid

Natterjack Toad

5.25 Development has a direct impact on biodiversity when it damages or destroys valuable wildlife habitats. Sustainable development, particularly within the urban environment, requires balancing human activities with protection of the environment. It is Government policy to make effective use of derelict and under-used land in urban areas. Essential development work, such as new housing, work places and transport-infrastructure, should take into account urban open spaces which have value as wildlife habitats.

5.26 There are two key issues to address:-
- how to ensure that development does not adversely affect environmental resources, so that where new development must take place, loss of biodiversity is avoided, reduced to a practical minimum or reversed; and
- how to enhance biodiversity in existing open spaces and in new development.

The Coastal Zone

5.27 The UK has the most diverse coastline in Europe, ranging from the rocky coasts of the south west peninsula and south west Wales to the chalk coasts of south east England, the low-lying soft coasts of East Anglia, the sea lochs of western Scotland and the basalt and limestone cliffs of Northern Ireland.

5.28 The extent and variety of the UK coastline has resulted in rich assemblages of plant and animal species, and many habitats of international importance. Our estuaries are especially significant, since these form over a quarter of the estuarine resource of western Europe. Internationally, the UK has a particular responsibility for conserving its estuaries for their bird populations, though they are also important for their prolific marine plant and invertebrate life, as fish spawning and nursery grounds, and for populations of larger animals such as seal and otters. Rock coastlines, while not so diverse as soft coasts, also support many species rarely seen elsewhere in Europe.

5.29 There are a number of threats to biodiversity within the coastal zone. These include:-
- pressures on coastal habitats arising from built development (including coastal engineering projects such as sea defences and barrages), recreation and tourism;
- rising sea levels, which taken together with coastal squeeze are contributing to a significant loss of coastal habitat; and
- pollution and contamination, including nutrient enrichment from sewage and fertiliser run-off, chemicals, oils etc.

Marine

5.30 Some 70% of the earth's surface is covered by oceans, which sustain a greater variety of life forms than found on land, but probably fewer numbers of species. It is often mistakenly believed that these ecosystems are very isolated from man and his economic activities: in fact, it is in the oceans where some of the issues of sustainable use of biodiversity come into sharpest focus. The direct exploitation of marine biodiversity in commercial fishing presents a major challenge to our ability to use natural resources in ways which sustain the target species, their ecosystems and the economic activity that depends upon them.

5.31 The quality of the marine environment around the UK is affected by human activities both on land and at sea, and natural factors such as tides, currents and the weather. Human activities on land result in the discharge of effluents containing metals, nutrients, organic compounds which reach the sea via rivers and estuaries, directly from the coast and from the atmosphere.

5.32 Shipping can present a risk of oil spills, and both shipping and oil and gas exploration have operational discharges which may cause potentially harmful concentrations and depositions which could adversely affect marine life.

5.33 The key issues are:-
- maintaining and improving current controls on man-made inputs to the sea, particularly chemicals which are toxic, persistent and liable to accumulate in living things;
- continuing pressure on many fish stocks which can have adverse effects on the food supply of other species, particularly other fish;
- the extent to which certain fishing practices lead to casualties among non-target species particularly seabed communities and birds, seals and small cetaceans, ie dolphins.

Recreation and Tourism

5.34 Much of our countryside is of great beauty, and is a focus for recreation and tourism. For example, our National Parks receive well over 50 million each year, and overseas visitors are attracted as much by our varied landscapes as by our history and culture.

5.35 Although the main threats to biodiversity lie elsewhere, excessive visitor pressures can be damaging at heavily used sites, including wear and tear on some habitats from trampling, disturbance to moorland birds reducing breeding success at some sites, and disturbance to wildfowl in areas used for water sports. It is therefore important that noisy, disruptive and damaging activities are carefully managed.

Coastal Sand Dunes

Opportunities for Enhancing Biodiversity

5.36 The box opposite gives examples of the ways policies can tackle these problems.

OPPORTUNITIES FOR ENHANCING BIODIVERSITY

Air Quality
- Reduce airborne emissions from power generation, industrial combustion processes, domestic heating and road transport.
- Pursue international agreements to ensure that other countries do the same.

Water
- To control, as far as possible, contamination from diffuse sources.
- To manage discharges of waste water.
- Specific new water resource developments:-
  - effective demand management;
  - assessment of groundwater vulnerability and groundwater protection;
  - appropriate augmentation schemes;
  - encourage greater variety of riverside vegetation and restoration of bankside habitats through more sympathetic water course management and farming practices.

Energy
- Support renewable energy through the Non-Fossil Fuel Obligation.
- Encouragement of energy efficiency and conservation.
- Research into cleaner forms of coal and oil fuel including the environmental impact of alternative methods.

Transport
- Encourage development where people live close to work and shops.
- Technological improvements to reduce the effect of emissions.
- Encourage forms of transport such as walking and cycling, and alternative forms of transport to the motor vehicle.

Farming
- Encourage farming and land management practices that enhance natural and locally characteristic diversity of flora and fauna.
- Improve livestock management to minimise pollution from wastes.
- Establish stocking densities and practices on moors, heaths and semi-natural grasslands in keeping with the environmental carrying capacity of the land.
- Improve crop management to minimise the use of fertilisers and pesticides.
- Encourage technological and other innovation to develop environmentally sympathetic farming methods.
- Encourage the commercial use of crop varieties and traditional breeds which are particularly adapted to the specific environment of a region.
- Recognise the importance of those traditional skills and practices upon which many valued habitats depend.
- Encourage greater diversity on the farm, for example through the encouragement of reversion of arable land to pastoral use in appropriate areas and the use of more varied rotations.
- Maintain, restore or reinstate hedges where possible and appropriate.

Woodlands
- Create new woodlands, especially in areas of low wildlife value.
- Conserve and restore as appropriate the remaining areas of ancient and semi-natural woodland.
- Encourage the sensitive management of other existing forests and woodlands including the restructuring of even-aged plantations to introduce the diversity of species and age classes, the use of silvicultural techniques that seek to mimic natural processes, the incorporation of open space, the retention of old trees and dead wood, improved management of rides and riparian zones, the creation of woodland edge habitats and the removal of invasive species.
- Encourage the use of native species of local provenance and where possible their establishment through natural regeneration.
- Promote the use of good quality local genetic hardwood stock where native broadleaved tree species are being planted and preserve local genotypes through the careful selection of seed sources.

Urban
- Create new urban "wildspace" within existing built up areas through more imaginative management of parks, other public open spaces and road verges.
- Encourage householders to enhance the biodiversity of their own gardens.

Coastal Zone
- The adoption of soft engineering approaches to coastal defence, including in appropriate cases, setting back the line of defences.
- Encourage the development of estuary and other local management plans.
- Encourage the co-ordinated efforts of those organisations responsible for managing inshore waters, the shoreline and the coastal fringe.

Marine
- Integrate environmental concerns into fisheries policies.
- Balance fishing effort against the natural ability of fish stocks to regenerate.
- Undertake more research into the full environmental impact of fishery policies and practices on other species and the wider environment, and seek ways to minimise adverse effects.
- Further reduce the discharge of contaminants such as heavy metals and PCBs into the sea either directly, or via input to rivers and to the atmosphere.

PARTICULAR ISSUES

5.37 The previous section has looked at generic themes and issues, and opportunities for current and future Government policy to assist the conservation, and where practicable, the enhancement of biodiversity. We have also carried out an analysis of the species and habitat action plans and the habitat statements (although we recognise the latter do not contain finalised proposals for conservation measures).

5.38 It was evident to us that there were a number of common issues which emerged from the analysis of key habitats and species. These issues, if tackled, would in our view make a major contribution to sustaining and enhancing biodiversity. Our analysis underlines the case for continued and vigorous implementation of existing measures, and further development of the many policies which are beginning to incorporate biodiversity objectives. In some areas, a change of direction may be required.

Scots Pine Forest

5.39 What struck us was the fact that many of our conclusions, reached from the bottom-up analysis, converged with the top-down view of these issues, and in many cases our concerns reinforce those reached elsewhere. To take one example, the Ministry of Agriculture, Fisheries and Food CAP Review Group concludes, with the support of the present Minister, that radical reforms of the CAP are needed. The Report notes that agricultural policy should recognise its impact on the environment, including wildlife habitats and species diversity, by setting clear objectives for environmental conservation and management and giving achievement of these a much higher priority than at present.

5.40 The Report also notes that the impact of the CAP on the environment is difficult to assess precisely, but reduction in production-related support for agriculture should have a number of positive benefits for the environment. The Report goes on to state that within the overall concept of sustainable agriculture, key environmental aims would cover preservation of species and habitats, quality and depth of soil, quality of water and the maintenance of desired landscape features. Our work serves to underline the importance of these conclusions.

Analysis of Species Action Plans

5.41 From the specific threats and conservation measures found to be common to a significant number of individual action plans, 12 principal threats and 13 conservation measures were identified. These are shown in Table 4. The 12 threats are those most prevalent among the 116 species for which action plans are published with this Report, and the 13 conservation measures are those that, if implemented, would address these threats.

5.42 When interpreting this information, three points should be borne in mind. Although many more references are made in the plans to some measures (for example better advice for land managers or others with an interest in the species are identified in 78 plans, but sensitive development planning and control in only five plans), nevertheless, we consider all the conservation measures identified to be of importance, and no order of priority is intended. Second, because there are very few marine species in the list, marine issues do not feature strongly. Third, some habitats are species-rich (eg. chalk grassland) while others are naturally species-poor (eg heaths). Analyses based on numbers of species are therefore more likely to indicate issues affecting the former rather than the latter habitat types.

5.43 The conservation measures identified are:-
- **Provide better advice for land managers or others with an interest in the species.**
  Land managers are sometimes unaware of the presence of rare or declining species on their land, or how the land may be managed to benefit the species. Often all that is required is the provision of advice rather than financial support. Provision of advice may also be usefully extended to those with other interests in the species concerned, or the land containing them, for example local conservation groups or land users.
- **Undertake ecological research, including distribution and reasons for decline.**
  We do not yet know why many species are declining, so several plans call for ecological research on habitat requirements, population dynamics, genetic viability and other factors which may affect the survival of the species. The need for this sort of research is often linked to the following measure.

- **Determine conservation or taxonomic status.**

  Where the plans show it is possible that some rare and little known species may be more common than recent records actually suggest, there is a need for systematic surveys to attempt to determine range and population size or the precise taxonomic status of a species.

- **Provide improved site protection.**

  Many action plans call for existing habitat sites, additional known sites, or any sites discovered in the future to be given some protection. Site protection is not always the most effective measure required to combat a known threat, but is often an important contributory factor.

- **Undertake re-introduction.**

  Rare species would benefit from re-introduction to former sites, or translocation to new ones - once the habitat conditions are right - to ensure their long term survival.

- **Improve or maintain water quality or quantity.**

  Most of the freshwater species concerned require good quality water to breed successfully, but some are believed to be suffering from the effects of pollution. Fertiliser run-off leading to nutrient enrichment is a problem. The plans recommend that efforts are made to improve water quality through the achievement of appropriate water quality objectives. Allied to the requirement for good water quality is the need for it to be available in suitable quantities, and for the water table to be maintained. As a result, this conservation measure is often linked with the following.

- **Undertake wetland or pond restoration.**

  The loss of wetland, through drainage or water abstraction, is a limiting factor for a number of the plants and animals. Some species require restoration of their wetland habitat if their long term survival in the UK is to be guaranteed.

- **Improve grassland management.**

  Many lowland species suffer from too little grazing of their habitat by livestock. Examples of such species are the early gentian, the marsh fritillary and the silver-spotted skipper. Conversely, other species like the skylark, corncrake and brown hare would benefit from extensification of grassland use. This dichotomy highlights one of the difficulties in conserving biodiversity in the natural environment. Some threatened plants and animals would benefit from increased financial support for grassland management, or from changes to the CAP. For other species, advice or encouragement to land managers to permit a level of grazing suitable for the species may suffice.

- **Enforce existing legislation more rigorously.**

  A few species, although protected under the WCA 1981, are still threatened by specimen collection and require rigorous enforcement of the law to ensure their survival.

- **Increase public awareness.**

  Some species are declining partly because people are not aware either that they are endangered or that certain activities are damaging to them. This applies to special interest groups too, for example, providing appropriate advice and guidance to rock climbers could help to conserve the Derbyshire feather moss and the wild cotoncaster by avoiding the rock faces where they grow.

- **Improve or encourage management schemes for woodland.**

  Woodland neglect or inappropriate management are major problems for a few species, in particular those which need either an open woodland floor, or temporary clearings as created through coppice management. Different management techniques are needed for which land managers may need financial assistance.

- **Review agro-chemical use on farmland; also, combat pesticide and fertiliser drift from adjacent land, and air pollution caused by vehicle emissions.**

  The causes of the rapid decline of several widespread farmland species such as the skylark, grey partridge, song thrush and pipistrelle bat are not fully understood. Research suggests that numbers may have been affected by the increased use of agro-chemicals. It would be desirable to assess the effect of these chemicals, and to find management procedures that would allow threatened species to recover without unduly lessening the productivity of the land. Airborne chemical pollution has also been identified as a threat to five species. Development of environmentally sensitive products and techniques is a scientific and technological challenge.

- **Provide more sensitive development planning and control.**

  Populations, or certain local populations of a small number of species are under either actual or potential threat from activities which could be controlled by local authorities through development planning or control processes. The dormouse is included in this category as road developments disrupt its wildlife corridors. Other species include the Norfolk flapwort (threatened by a road-widening scheme), Shetland pondweed (a housing development), and a fungus (Tulostoma niveum) which could be

Yellow Marsh Saxifrage

eliminated by road widening. Although no other species are specifically identified in the action plans as being under such immediate threat, many plans call for the interests of threatened species to be taken into account in local development plans. It should be emphasised that local authorities have a key role to play in the protection of species in their areas. Planning Policy Guidance on Nature Conservation (PPG9) provides comprehensive advice to local authorities in England on how the Government's policies for nature conservation are to be reflected in land-use planning under domestic and international law. Local planning authorities must take PPG9 into account in preparing development plans, and it may be material to decisions on individual planning applications.

5.44 We recognise that the species plans analysed do not reflect the needs of UK biodiversity as a whole. This is because the species were not selected on an ecosystem basis, but according to other criteria. Marine species are poorly represented. The parallel analysis of issues within habitat action plans (see next section) should help to redress any such imbalance. There are some threats and associated conservation measures which apply to one, or a limited number of, species because they are specialised in their conservation needs (especially habitat requirements). Because this analysis identifies issues common to a significant number of action plans, it does not encompass such unique threats and measures, but their importance should not be underestimated.

P Wakely

Heath Fritillary

Table 4: SPECIES ACTION PLANS: PRINCIPAL COMMON ISSUES

Threats		Conservation measures needed	
Type	**Main causes**	**Measure**	**Action (number of species expected to benefit)**
Habitat loss and fragmentation	Change of grassland to arable Built development Coniferisation of woods Intensification of farming Land drainage and water abstraction	Habitat protection, re-creation and connection Extensification of farming	Protected areas (64) Re-introduction of species (51) Improved or maintained water quality or quantity (32) Improved grassland management (23) Land management schemes for woodland (8) More sensitive development planning and control (5)
Restricted distribution or falling population despite apparent availability of habitat	Largely unknown	Research and survey	Ecological research (75) Determine true status (52)
Natural succession or competition from other native species	Lack of grazing	Improved management of existing sites	Better advice for land managers (78) Improved grassland management (23)
Habitat quality decline	Fertiliser run-off Falling water quality Insecticides affecting birds and bats	Extensification of farming and further control of pollution in freshwater habitats	Improve water quality (32) Review of insecticide use on farmland (11)
Direct human disturbance to species	Specimen collection Public access	Improved species protection	Enforce existing legislation more rigorously (13) Increase public awareness (12) Sensitive development planning and control (5)

Table 5: EXAMPLES OF SPECIES LIKELY TO BENEFIT FROM CONSERVATION MEASURES IDENTIFIED

Conservation measure	Examples of species likely to benefit		
	Plants and fungi	**Invertebrates**	**Vertebrates**
Provide better advice for land managers or others	three-lobed crowfoot floating water plantain Killarney fern	netted carpet moth speckled footman moth medicinal leech shining ram's-horn	grey partridge capercaillie natterjack toad otter red squirrel
Undertake ecological research, including distribution and reasons for decline	Lundy cabbage marsh earwort Derbyshire feather moss	freshwater pea mussel speckled footman moth bright wave moth shrill carder bee	song thrush skylark pollan dormouse harbour porpoise red squirrel
Determine conservation or taxonomic status	Newman's lady fern green shield-moss wild cotoneaster eyebrights	black-backed meadow ant blue ground beetle	aquatic warbler song thrush pipistrelle bat water vole
Provide more site protection	Devil's bolete lady's slipper orchid orange-fruited elm lichen marsh earwort Young's helleborine eyebrights petalwort	white-clawed crayfish medicinal leech silver spotted skipper butterfly netted carpet moth southern damselfly	great crested newt greater horseshoe bat Scottish crossbill aquatic warbler
Undertake re-introduction	fen orchid lady's slipper orchid Norfolk flapwort creeping marshwort	heath fritillary large blue butterfly black-backed meadow ant narrow-headed ant starlet sea anemone	great crested newt otter pollan sand lizard natterjack toad
Improve or maintain water quality or quantity	fen orchid holly-leaved naiad slender naiad river jelly lichen	shining ram's-horn fresh-water pearl mussel glutinous snail	otter twaite shad allis shad vendace
Undertake wetland or pond restoration	petalwort fen orchid ribbon-leaved water plantain starfruit	black bog ant southern damselfly sandbowl snail medicinal leech	bittern great crested newt
Improve grassland management	orange-fruited elm lichen early gentian nail fungus yellow marsh saxifrage	shrill carder bee marsh fritillary hornet robberfly mole cricket silver-spotted skipper high brown fritillary	corncrake skylark brown hare greater horseshoe bat
Enforce existing legislation more rigorously	Killarney fern Derbyshire feather-moss elm gyalecta	white-clawed crayfish freshwater pearl mussel	water vole (song thrush, stone curlew - through the EU)
Increase public awareness	holly-leaved naiad lady's slipper orchid Lundy cabbage	stag beetle freshwater pearl mussel shrill carder bee hornet robberfly	pipistrelle bat corncrake
Improve or encourage land management schemes for woodland	*Schismatomma graphidioides* *Lejeunea mandonii*	blue ground beetle high brown fritillary pearl-bordered fritillary	capercaillie dormouse greater horseshoe bat
Review insecticide use on farmland and combat other forms of airborne pollution	elm gyalecta marsh earwort	hornet robberfly mole cricket	skylark grey partridge song thrush pipistrelle bat
Provide more sensitive development planning and control	Shetland pondweed Norfolk flapwort *Tulostoma niveum*	reed marsh beetle	dormouse

Analysis of Habitat Action Plans and Habitat Statements

5.45 The following issues and actions apply **across the range of UK habitats**:-

- the habitat action plans and directional statements identify the need for protection and management of important UK habitats;
- the plans and statements emphasise the need to secure appropriate management of habitats, including positive action by landowners or managers to benefit the conservation of species contained in the habitat;
- there is a need for bodies with statutory responsibilities for planning, development, land management or conservation to take into account the Government's policies on sustainable development and the conservation of biodiversity; and
- pollution, in particular deposition of acidifying compounds and nutrients, is a major concern.

Saltmarsh Solway Firth

5.46 The following issues and actions are common to **all terrestrial (and some coastal) habitats**:-

- the plans and statements call upon authorities throughout the UK to ensure that policies for the protection of designated areas are reflected in all appropriate local and development plans, or indicative forestry strategies;
- the conservation of local and regional character is important. Work being undertaken by the Countryside Commission and the country agencies will assist all relevant authorities in promoting this;
- the action plans and statements identify that an underlying threat to many habitats is the pressure of agricultural production encouraged by the Common Agricultural Policy; and
- environmental land management schemes are cited as necessary for habitat management. More use of schemes such as Countryside Stewardship, ESAs and the Woodland Grant Scheme is called for, together with continual refinement of such initiatives to meet habitat needs.

5.47 The following issues and actions apply to **marine (and some coastal) habitats**:-

- lack of protection and management of some habitats below the tideline is identified as a problem. The ability to declare Marine Nature Reserves and "no take zones" more easily is cited as a measure which could be of particular benefit;
- marine pollutants have been identified as a matter for concern. The Government is urged to build on the positive contribution to improving maritime safety and preventing marine pollution made by Lord Donaldson's report *Safer Ships, Cleaner Seas* following the 1993 Shetland oil spill. Where appropriate, further routes for tankers and other shipping are also desirable;
- updated planning guidance on coastal development should be introduced for Scotland and Northern Ireland to complement that which exists for England Wales; and
- other action identified in the plans was implementation of improved fishery management under the Common Fisheries Policy. The UK and other North Sea countries have asked the Commission to bring forward proposals to improve fisheries management. Progress will be reviewed at a meeting of Environment and Fisheries Ministers in March 1997.

5.48 The following issues and actions apply to **wooded habitats**:-

- further action is called for to encourage coppice management and to control the spread of non-native species;
- further woodland management guidelines, aimed specifically at the conservation of biodiversity, would be helpful; and
- the poor market in domestic timber products is a problem because it means there is often little profit in managing

broadleaved woodlands (this also indirectly encourages the import of tropical hardwoods which may be grown in an unsustainable fashion).

Rhos/Culm Pastures

5.49 The following issues and actions apply to **grazed or grassland and heath habitats**:-

- several reforms of EC agricultural policy are called for in the plans and statements, these include reform of livestock subsidies and quotas to reduce over-grazing in uplands and allow grazing to return to some lowland habitats; and
- the plans and statements also call for action to overcome the problems of managing common land.

5.50 The following issues and actions apply to **cultivated farmland**:-

- there is a need to consider ways of encouraging more extensive farming including the promotion of cereal field margins; and
- the improved conservation of hedgerows, isolated trees and farm ponds would make a significant contribution to maintaining biodiversity. This could be promoted through farm advisory services, with advisers trained in habitat management for biodiversity.

5.51 The following issues and actions apply to **freshwater and bog habitats**:-

- habitat fragmentation, leaving small populations vulnerable to extinction and leading to genetic isolation, is identified as a particular problem for wetland habitats. The plans call for further restoration of these habitats through appropriate land management schemes;
- falling water tables, due to ground water or surface water abstraction is another major issue requiring the further production and implementation of water level management plans. A review of water abstraction licences would also be helpful in problem wetland sites;

- water pollution, especially nutrient enrichment, is a further major problem. The introduction of Statutory Water Quality Objectives, as appropriate, (subject to successful tests in pilot catchments) could help address this. In the meantime, the non-statutory River Quality Objectives and Statutory Urban Waste Water Treatment (91/271/EEC) Directive standards provide the framework for improving water quality. The Directive requires nutrient reduction treatment by December 1998 for large sewage treatment works (STWs) discharging into waters identified or Sensitive Areas (Eutrophic). The Government has designated 33 such areas (as freshwaters) and phosphorous removal at some 41 STWs will be required. Reductions in nitrates leaching into water from agriculture will result from the introduction of the Nitrate Directive. In addition, the proposed Directive on Ecological Quality of Surface Waters may also help to address the problem of eutrophication; and
- the need to develop improved training and advisory services on technical management has also been identified.

Liverworts

Table 6: HABITATS: PRINCIPAL COMMON ISSUES

Threats	Conservation measures needed
All UK habitats	
Inadequate or inappropriate site management.	Provide advice to land managers where appropriate or ensure that advice and information is available. Conservation agencies should negotiate voluntary management agreements or, as a last resort, use the compulsory purchase powers contained in the Conservation Regulations 1994.
Lack of requirement for statutory bodies to have regard to Government policies on sustainable development.	Clarify the responsibilities of relevant bodies.
Airborne pollution.	Identify critical loads and levels for key pollutants, such as sulphur and nitrogen oxides, and introduce measures to reduce emissions. Introduce measures to reduce emission of greenhouse gases in line with decisions taken during the Berlin mandate negotiations.
All terrestrial (and some coastal) habitats	
Habitat destruction, or lack of suitable management.	Improve habitat protection within development, mineral, waste disposal and coastal zone management plans, and indicative forestry strategies. Improve planning and management of recreation. Discourage or prohibit inappropriate types, or levels of use. Further use and refinement of land management schemes such as ESAs, Countryside Stewardship, Tir Cymen and Woodland Grant Scheme.
Over-grazing in upland woods and moors, and under-grazing on many lowland habitats.	Provide advice on appropriate management regimes to land managers. Use existing measures such as cross compliance, Woodland Grant Scheme, ESAs, and the Moorland Scheme to encourage reduction of grazing where necessary. Seek reform of the relevant parts of CAP.
Marine (and some coastal) habitats	
Lack of protection and management of the sea below the tide line.	Encourage inter-agency co-operation to conserve habitats below low water mark. Simplify the designation of protected areas (eg: MNRs or "no take zones").
Marine pollution and contamination, especially by oil, persistent and bio-accumulating chemicals, and nutrients.	Where appropriate introduce statutory water quality objectives. Consider making hazard assessment mandatory within marine chemical licensing procedures. Speed up efforts to phase out or ban substances such as PCBs and organo-halogens. Consider introducing further routes for tankers and other shipping to avoid sensitive areas. Liaise with the IMO to minimise the impact of shipping.
Over-fishing and damaging fishing techniques.	Seek to improve fishery management under the Common Fisheries Policy, including assessment of potential effects of fisheries on non-target species and the environment. Minimise damage to seabed habitats and communities by providing advice on use of fishing gear and developing environmentally-friendly gear.

Wooded habitats

Insufficient woodland management.	Continue refinement of broadleaved woodland incentives, including those for coppice management and ancient semi-natural woodland restoration.
Lack of guidance on how to maintain and enhance biodiversity in woodland.	Production and promotion of further relevant guidelines.
Lack of markets for domestic products, reducing the incentive to manage UK woodlands and indirectly furthering the destruction of tropical forests.	Work with World Trade Organisation to clarify and, if necessary, extend environmental trade restrictions, considering restricting imports of timber from unsustainable sources if necessary.

Grazed grassland and heathland habitats

Overgrazing in uplands and undergrazing of semi-natural habitats in lowlands.	Provide advice on appropriate management regimes to land managers. Seek to reform livestock subsidy and quota systems.
Unsympathetic management of common land.	Re-consider introduction of legislation to facilitate the favourable management of common land.

Cultivated farmland

Lack of protection and management of hedges and other boundary features, and isolated trees and ponds.	Using existing measures (eg: Countryside Stewardship; hedgerow schemes or, if necessary, attach relevant management conditions to Arable Area Payments) to encourage the appropriate management of boundary features. Extend boundary features, connecting isolated habitats where possible.
Intensive cropping regimes and use of agrochemicals.	Encourage extensification. Encourage more targeted use of agrochemicals.
Insufficient advice to farmers on how to manage arable land to conserve biodiversity.	Improve the advice offered to farmers.

Freshwater and bog habitats

Habitat fragmentation leading to small, vulnerable populations and genetic isolation.	Promote the restoration of wetland habitats through the further use of appropriate schemes.
Falling water tables.	Continue to produce and implement water level management plans and review water abstraction licensing procedures.
Water pollution, especially nutrient enrichment.	Test the implementation and operation of statutory water quality objectives, where appropriate, especially for phosphates.
Lack of technical advice on the management of wetlands for wildlife.	Develop appropriate training and advisory services.

Coastal habitats

Aggregate extraction.	Review current licences for aggregate extraction, and consider modifying those where impact on biodiversity is likely to be severe. Avoid issuing licences where these will have direct, or indirect effects on sites of conservation importance.
Sea level rise threatening saltmarsh and other fringing habitats.	Prepare a strategic response to sea level rise which would permit protection or creation of important habitats. Research methods of estuarine habitat management and the creation of inter-tidal habitats.
Coastal planning guidance in Scotland and Northern Ireland needs updating.	Update appropriate Planning Guidance Notes.

SECTION 4

CHAPTER 6

THE WAY FORWARD

THE TASKS

6.1 The Steering Group had four tasks:

- to recommend targets for key species and habitats, and to cost them;
- to suggest ways of improving standardisation and access to information on biodiversity;
- to recommend ways of increasing public awareness and involvement in conserving biodiversity; and
- to recommend ways of ensuring that commitments in the Plan are properly monitored and carried out.

Our proposals are summarised in this chapter. Our remit was largely the territory of the UK and Northern Ireland. But, in view of the importance of biodiversity in the UK Crown and Dependent Territories, we asked the authorities concerned to provide a report on progress towards implementing the Biodiversity Convention. This is shown at Annex B.

TARGETS FOR KEY SPECIES AND HABITATS

6.2 Our remit was to develop a range of specific costed targets for key species and habitats for the years 2000 and 2010, and to publish these in 1995. Step 33 on page 167 of Biodiversity: The UK Action Plan committed the Government to:

> *"Prepare action plans for threatened species in priority order: globally threatened, threatened endemics; other threatened species listed in the relevant schedules and annexes to UK and EC legislation and international agreements to which the UK is a party; endangered and vulnerable species listed in Red Data Books, aiming to complete and put into implementation plans for at least 90% of the presently known globally threatened and threatened endemic species within the next ten years".*

6.3 Because of the unsystematic way in which international obligations have been defined over time, the Steering Group thought it appropriate to revisit the criteria and decided to include, in addition, costed action plans for rapidly declining species (the populations of which have declined by more than 50% in the last 25 years), particularly those where there is high public interest, on the grounds that action needs to be taken now if we are to discharge our conservation obligations under the Biodiversity Convention and other international agreements.

Costing the Plans

6.4 Best endeavours were made to estimate costs precisely, though this is an inexact science. For this purpose, targets have been set although we acknowledge that in some cases it was not possible to do so with great precision. Costs are therefore indicative only.

6.5 Table 7 provides indicative costs for each Species Plan for the years 1997, 2000 and 2010. Note that the costs, for the UK as a whole,

- are additional to existing financial commitments;
- are net of any savings or income;
- make no assumptions about where the costs might lie. On average, half of the cost of action plans now in hand has been borne by the non-Government sector;
- to avoid double counting we have only included habitat management, restoration and creation costs and water quality or quantity improvement costs where the habitat is not a key one (and so is not subject to a separate action plan) or, (when it is), where the conservation action required is not normal for that habitat; and
- 10% has been added to the grand total to cover administration.

6.6 The checklist of action points, which was used as appropriate to cost each plan, includes:-

- survey(s) to determine distribution and population size;
- ecological research;
- genetic and population dynamics studies;
- ex-situ conservation (cultivation and captive breeding);
- seed bank creation and maintenance;
- re-introduction and translocation;
- special habitat management and restoration;
- habitat creation;
- any special new land management scheme that may be needed;
- control of competitors and predators;
- wardening of sites;
- special water quality and quantity improvements;
- monitoring;
- advice to land managers; and
- public relations.

Shining Ram's Horn Snail

Table 7: INDICATIVE COSTS FOR SPECIES PLANS

Vertebrate species				
English name	**Latin name**	**1997** **(£000 per annum)**	**2000** **(£000 per annum)**	**2010** **(£000 per annum)**
Water vole	*Arvicola terrestris*	150	110	105
Brown hare	*Lepus europaeus*	50	50	25
Otter	*Lutra lutra*	105	90	70
Dormouse	*Muscardinus avellanarius*	35	60	35
Greater mouse-eared bat	*Myotis myotis*	1	1	1
Pipistrelle bat	*Pipistrellus pipistrellus*	20	10	5
Greater horseshoe bat	*Rhinolophus ferrum-equinum*	45	45	45
Red squirrel	*Sciurus vulgaris*	220	210	210
Harbour porpoise (+other small cetaceans)	*Phocoena phocoena*	400	250	250
Aquatic warbler	*Acrocephalus paludicola*	7	5	2
Skylark	*Alauda arvensis*	104	103	93
Bittern	*Botaurus stellaris*	10	10	10
Stone curlew	*Burhinus oedicnemus*	105	100	50
Corncrake	*Crex crex*	550	330	360
Scottish crossbill	*Loxia scotica*	40	20	20
Grey partridge	*Perdix perdix*	95	108	88
Capercaillie	*Tetrao urogallus*	90	30	20
Song thrush	*Turdus philomelos*	124	108	88
Sand lizard	*Lacerta agilis*	80	75	65
Great crested newt	*Triturus cristatus*	110	90	65
Natterjack toad	*Bufo calamita*	37	32	23
Allis shad	*Alosa alosa*	23	21	16
Twaite shad	*Alosa fallax*	31	29	20
Pollan	*Coregonus autumnalis*	14	20	14
Vendace	*Coregonus albula*	30	32	16
TOTAL VERTEBRATES		2476	1939	1696

Invertebrate species				
English name	**Latin name**	**1997** **(£000 per annum)**	**2000** **(£000 per annum)**	**2010** **(£000 per annum)**
High brown fritillary	*Argynnis adippe*	21	12	12
Pearl bordered fritillary	*Boloria euphrosyne*	18	12	12
Marsh fritillary	*Eurodryas aurinia*	20	6	6
Heath fritillary	*Mellicta athalia*	18	11	11
Large copper	*Lycaena dispar*	8	6	4
Large blue butterfly	*Maculinea arion*	47	47	35
Silver spotted skipper	*Hesperia comma*	9	9	9
Speckled footman moth	*Coscinia cribraria*	7	10	8
Netted carpet moth	*Eustroma reticulatum*	10	5	2
Bright wave moth	*Idaea ochrata*	6	5	3
Shrill carder bee	*Bombus sylvarum*	13	14	6
Black bog ant	*Formica candida*	13	9	4
Narrow-headed ant	*Formica exsecta*	14	5	4
Black-backed ant	*Formica pratensis*	9	6	4
Southern damselfly	*Coenagrion mercuriale*	21	13	12
Mole cricket	*Gryllotalpa gryllotalpa*	7	6	6
White-clawed crayfish	*Austropotamobius pallipes*	41	32	30
Medicinal leech	*Hirudo medicinalis*	17	10	9
Ivell's sea anemone	*Edwardsia ivelli*	2	1	1
Starlet sea anemone	*Nematostella vectensis*	3	5	3
Hornet robber fly	*Asilus crabroniformis*	39	31	23
A hoverfly	*Callicera spinolae*	6	3	3
A hoverfly	*Crysotoxum octomaculatum*	10	9	5
A dung beetle	*Aphodius niger*	4	3	3
A ground beetle	*Bembidion argenteolum*	3	2	1
Blue ground beetle	*Carabus intricatus*	7	4	2
A broad-nosed weevil	*Cathormiocerus brittanicus*	3	2	1
A leaf beetle	*Cryptocephalus coryli*	9	7	4
A leaf beetle	*Cryptocephalus exiguus*	1	0	0
Violet click beetle	*Limoniscus violaceus*	8	7	3
Stag beetle	*Lucanus cervus*	10	7	4
A long-horn beetle	*Obera oculata*	5	4	2
A ground beetle	*Panagaeus crux major*	6	4	3

English name	Latin name	1997 (£000 per annum)	2000 (£000 per annum)	2010 (£000 per annum)
Reed marsh beetle	Stenus palposus	5	3	3
A ground beetle	Tachys edmondsi	1	0	0
A snail	Anisus vorticulus	12	9	8
Sandbowl snail	Catinella arenaria	4	3	2
Fresh-water pearl mussel	Margaritifera margaritifera	41	13	10
Glutinous snail	Myxas glutinosa	7	1	1
Freshwater pea mussel	Pisidium tenuilineatum	8	7	5
Depressed river mussel	Pseudanodonta complanata	6	3	2
Shining Ram's horn snail	Segmentina nitida	12	9	8
Narrow-mouth whorl snail	Vertigo angustior	6	3	3
A whorl snail	Vertigo genesii	6	1	1
A whorl snail	Vertigo geyeri	7	3	3
Desmoulin's whorl snail	Vertigo moulinsiana	3	1	1
TOTAL FOR INVERTEBRATES		533	363	282

Plant species				
English name	Latin name	1997 (£000 per annum)	2000 (£000 per annum)	2010 (£000 per annum)
Ribbon-leaved water plantain	Alisma gramineum	6	3	2
Creeping marshwort	Apium repens	7	4	3
Norwegian mugwort	Artemisia norvegica	2	1	1
Newman's lady fern	Athyrium flexile	13	6	2
A fungus	Battarraea phalloides	5	2	1
Devil's bolete	Boletus satana	3	1	1
Starry breck lichen	Buellia asterella	1	4	3
Green shield moss	Buxbaumia viridis	11	8	3
Orange-fruited elm lichen	Caloplaca luteoalba	28	34	38
Mossy stonewort	Chara muscosa	11	12	8
Mountain scurvy grass	Cochlearia micacea	7	1	1
Lundy cabbage	Coincya wrightii	3	2	1
River jelly lichen	Collema dichotomum	17	22	9
Wild cotoneaster	Cotoneaster cambricus	2	1	1
Lady's slipper	Cyripedium calceolus	19	17	12
Starfruit	Damasonium alisma	4	4	2
Glaucous beard moss	Didymodon glaucus	2	1	1
Cornish path moss	Ditrichum cornubicum	3	2	1
Slender green feather moss	Drepanocladus vernicosus	2	6	7
Young's helleborine	Epipactis youngiana	3	1	1
Seven eyebright species	Euphrasia spp.	4	1	1
Western ramping fumitory	Fumaria occidentalis	7	12	6
Early gentian	Gentianella anglica	23	10	4
Elm gyalecta	Gyalecta ulmi	9	8	3
Marsh earwort	Jamesoniella undulifolia	16	13	7
Norfolk flapwort	Leiscolea rutheana	4	2	1
A liverwort	Lejeunea mandonii	14	12	6
Fen orchid	Liparis loeselii	19	7	5
Floating water plantain	Luronium natans	12	7	6
Western rustwort	Marsupella profunda	7	3	2
Holly-leaved naiad	Najas marina	25	13	3
Slender naiad	Najas flexilis	19	18	11
Petalwort	Petalophyllum ralfsii	19	16	6
Nail fungus	Poronia punctata	9	7	6
Shetland pondweed	Potamogeton rutilus	15	8	5
A lichen	Pseudocyphellaria aurata	1	0	0
A lichen	Pseudocyphellaria norvegica	12	13	7
Three-lobed water crowfoot	Ranunculus tripartitus	5	7	3
Shore dock	Rumex rupestris	14	9	4
Yellow mountain saxifrage	Saxifraga hirculus	7	2	2
A lichen	Schismatomma graphidioides	15	15	15
Derbyshire feather moss	Thamnobryum angustifolium	5	3	3
Killarney fern	Trichomanes speciosum	33	14	12
A fungus	Tulostoma niveum	4	2	1
A moss	Weissia multicapsularis	7	4	3
TOTAL FOR PLANTS		454	338	220
GRAND TOTAL (includes 10% additional costs to cover administration)		3809	2904	2418

Estimated Costs for Habitat Action Plans

6.7 A summary of the estimated costs for each habitat action plan in 1997, 2000 and 2010 is shown in Table 8. This covers public expenditure costs only. It excludes private sector resources (ie those private sector costs not reimbursed by grants). The costs are additional to existing (1995) public expenditure commitments, which is estimated at about £22.5 million in 1995/96.

6.8 It should be noted that the summary costs are best approximations due to the lack of accurate data in many cases. For example, several scheme payments are made on the basis of a specified area, and there are often two or more habitats covered within that area. The summary costs:-

- include the costs of managing public sector land, and the costs of land management scheme payments to land managers (including administration);
- take account of revenue from land management (eg reeds and grazing);
- include land purchase costs, both the costs of public sector acquisition and grants for private sector purchase; and
- exclude (unless stated) the costs of research, monitoring, advice to managers, site safeguard and designations and publicity, although these costs are likely to be small relative to habitat management costs and some of them should only appear in the Species Action Plan costings (where they relate to specific species) to avoid double counting.

6.9 To put these costs (and the Species Action Plan costs) in context they should be considered against the background of current expenditure commitments. In 1995/96 planned public expenditure provision for payments under agri-environment schemes in the UK is about £100 million. This includes spending on ESAs, Countryside Stewardship, Nitrate Sensitive Areas and Tir Cymen. It excludes the cost of scheme administration and monitoring. This spending accounts for about 3% of the total UK agricultural support payments of over £3 billion in 1995/96 (direct payments to farmers plus market support through export refunds etc).

6.10 In addition to this agri-environment spending, there is the expenditure of the statutory nature conservation organisations. English Nature's planned spend in 1995/96 is £41 million, Countryside Council for Wales is £17 million and Scottish Natural Heritage is £40 million. Other public expenditure which is relevant to nature conservation includes that by the National Rivers Authority, the Forestry Commission on the Woodland Grant Scheme, and Forest Enterprise and the Agriculture Departments on the Farm Woodland Premium Scheme.

6.11 The contribution of private sector funding is also important. In 1991, environmental non-Government organisations spent about £80 million on conservation in England and Wales. Taking account of spending in Scotland and Northern Ireland, conservation probably now benefits from over £100 million of NGO expenditure annually.

Table 8: INDICATIVE COSTS FOR HABITAT PLANS

Habitat Type	1997 (£000 per annum)	2000 (£000 per annum)	2010 (£000 per annum)
Fens	40	70	70
Mesotrophic Lakes	170	350	350
Chalk Rivers	500	1,000	1,100
Reedbeds	190	380	540
Grazing Marsh	4,200	8,400	13,200
Cereal Field Margins	500	1,100	2,100
Upland Oakwood	3,400	6,800	11,600
Limestone Pavement	130	130	100
Lowland Heathland	1,100	2,100	3,400
Purple Moor Grass and Rush Pastures	160	310	510
Ancient and/or Species Rich Hedgerows	1,000	1,700	3,000
Saline Lagoons	800	1,500	600
Seagrass Beds	330	330	330
Native Pine Woodlands	350	350	260
TOTAL	12,870	24,520	37,160

Local Targets

6.12 We have emphasised the importance of developing Local Biodiversity Action Plans both in their own right - to develop a strategy for species and habitats of local significance - and to help to deliver national targets. Guidance for the development of Local Biodiversity Action Plans, which makes good use of the voluntary approach, and a diagram which shows the relationship between local and national targets and how these may be implemented is set out in Annex C.

6.13 We recognise that provision of this guidance is only the first step in the implementation of Local Biodiversity Action Plans. It needs to be seen in the context of a new target-led approach to conservation stemming from the UK Action Plan. Implementation will require more than just the guidance to be a success. It will require training, pilot projects, community workshops and other means. Full consultation will be necessary with national bodies, especially those representing Local Authorities, and land managers so that the whole programme can be carried forward effectively.

6.14 This new initiative will require considerable care in the way that it is promoted with farmers and other land managers. The process described in the guidelines is a new mechanism which will present considerable challenges for all involved in creating effective voluntary partnerships between the community and land managers - whether in town or country. Participation of land managers through all stages of the process will be crucial to success.

6.15 The extent of community participation needed, hitherto rare for local authorities, is now becoming more widely accepted as an integral part of the Agenda 21 process. Production of Local Biodiversity Action Plans will help to inform those involved in the planning process in developing their work on biodiversity conservation.

6.16 Although a few examples of Local Biodiversity Action Plans already exist, none of these has involved the local community, particularly land managers and local businesses and industry, in the way advocated in the guidelines. We therefore suggest a series of pilot projects to provide examples of good practice including ways of involving the local community.

6.17 Experience gained in the implementation of Local Agenda 21 suggests that training for all participants in the process will be crucial to success. We consider that training should be given a high priority as an essential part of the implementation process for Local Biodiversity Action Plans.

6.18 We recognise the need for national co-ordination to promote consistency of approach, and to recommend standards of good practice in the production and implementation of Local Biodiversity Action Plans; also to provide guidance for development of pilot projects and the necessary training involved.

6.19 We recommend that a small national Advisory Group be established for this purpose, involving local Government, statutory agencies, national voluntary bodies, land managers and other relevant sectors. In view of the close relationship between the process recommended for development of Local Biodiversity Action Plans and the Local Agenda 21 process, it would seem to us appropriate for this Advisory Group to be a part of the UK Local Agenda 21 Steering Group, representing as it does all the Local Authority Associations. We are pleased that the Steering Group for Local Agenda 21 has agreed in principle to this course of action.

DATA AND INFORMATION

6.20 In Chapter 3, we recommend a three pronged approach to improving the quality and accessibility of data and biological recording. This is:-
- making the maximum use of existing data;
- developing a nationally based biodiversity information system by stages; and
- developing a locally based biodiversity information system through the establishment of local consortium funding.

6.21 The sharing and re-use of data should bring economies to national and local bodies as well as providing improved information to support the biodiversity targets. Moving towards a United Kingdom Biodiversity Database (UKBD) will require a staged approach, establishing a co-operative network with a small management centre and investing in products and standards which would encourage data owning organisations to improve the accessibility of their data. The need to encourage the development of Local Biodiversity Action Plans which will help to deliver both national and local targets is fundamental. The consortium approach, whereby local data centres would receive funds for providing a service to a range of bodies (eg local authorities, country agencies, NRA, wildlife trusts, RSPB regions etc) is considered to be the best way forward, while bearing in mind the need for a mechanism to co-ordinate and bring together local effort which is inevitably fragmented.

L Gill

Raised Bog

6.22 If all the work identified at both national and local levels were to be tackled simultaneously, and quickly, the cost would be very high and there would be a significant management overhead. We have therefore looked at a more realistic approach through the appropriate grouping and phasing of projects.

6.23 Best endeavours were made to develop indicative costs for the stages within the national and local approaches. Existing programmes will contribute to the process, but new investment and improved co-ordination are needed. The costs summarised below are based on the suite of more detailed projects contained within each stage. The costs include:-
- new investment to achieve the stage outputs; and
- the costs of programme and project management calculated at marginal costs for organisations which are already involved in relevant activities.

6.24 One option is to continue with the current level of investment. This has a number of risks. At the national level there is a risk that some datasets will be lost, or severely constrained because their value is not understood by their owners or their potential users, and without a co-ordinated strategy, general access and the collation of data will continue to be difficult. This will affect the ability to respond to international and European conservation initiatives, including the Habitats and Birds Directives, as well as the ability to monitor the proposed targets contained in this report.

Creeping Marshwort

6.25 At the local level investment in local records centres is declining and centres are closing. There is a significant risk that the development of Local Biodiversity Action Plans will be delayed or that they will be constrained in the longer term. We therefore recommend that appropriate investment be made at both the national and local levels.

6.28 For the national approach we see three stages:-

● identifying strategic datasets and setting standards;
● providing incentives to network members to improve accessibility;
● testing the benefits of an electronic network.

6.26 For the local approach there are also three stages:-
● establishing a consortium funding case and setting standards;
● supporting local initiatives and establishing or enhancing local data centres;
● building local systems and networks.

The National Approach

6.27 The first stage will be to identify strategic datasets and to set standards. The aim will be to provide:-
● a nationally agreed list of strategic datasets which underpin international, European and national requirements;
● standard lists for taxonomy, biotope and land use classifications, and a system for maintaining and distributing them as key information exchange standards;
● a maintained catalogue of biodiversity datasets containing information on their content and accessibility;
● a programme for publishing key datasets; and
● a framework for a data access agreement for adoption by bodies holding strategic national datasets.

The stage will need to be developed over a minimum of three years starting in 1996/7 and could be delivered for £575,000.

6.28 Stage two will be to provide incentives to network members to improve accessibility. This will be through the following products:-

● a series of improvements to the accessibility of datasets in the agreed list which will include cataloguing, documentation, dataset publishing, agreement on access terms and establishing a legal title;
● a data model to assist information exchange; and
● the collection of monitoring and surveillance data within the action plans for transfer to the appropriate national and local datasets.

Begun after the first stage has achieved consensus on the agreed list of strategic datasets, the second stage could be programmed over a minimum of three years and would cost £845,000.

6.29 Stage three will test the benefits of an electronic network. This stage will include:-

● a system for managing the selection of key habitats and species, and providing access to the organisations participating in the delivery of the action plan targets as an example of the long term data approach;
● a networked facility for providing the information exchange needed for each organisation co-ordinating either a habitat or a species action plan. The facility would be a pilot for wider networked access to data and information;
● a report on the best means of establishing an electronic network to make data access and exchange easier for the emerging network of organisations managing national or local biodiversity data; and
● a report looking at the benefits arising from national and local data management, and the added value from electronic networking.

Stage three could be delivered for £700,000 over three years and could beneficially overlap with stage two.

Great Crested Newt

6.30 The estimated costs for the projects contained in each stage are shown in Table 9.

Table 9: ESTIMATED COSTS FOR DEVELOPING A UK BIODIVERSITY DATABASE

Project	1996/7 (£000)	1997/8 (£000)	1998/9 (£000)	1999+ (£000)
Stage One				
Priority list of national datasets	50			
Establish and maintain standard term lists	75	25	25	
Establish and maintain data catalogue	50	50	50	
Design data exchange arrangements	15	10	10	
Publish datasets to make them accessible	80	60	20	
Management overhead	35	10	10	
TOTAL	305	155	115	
Stage Two				
Support establishing legal title		40	80	80
Support accessibility improvements		100	150	100
Establish and maintain generic data models		25	10	20
Support data banking		45	45	45
Management overhead		35	35	35
TOTAL		245	320	280
Stage Three				
Scoping study for electronic network	60	25		
Pilot of network and dataset access	195	95	95	
Key species and habitat selection system	45	25	25	
Evaluation of benefits			60	
Management overhead	25	25	25	
TOTAL	325	170	205	
GRAND TOTAL	630	570	640	280

The Local Approach

6.31 The first stage will be to devise a case for consortium funding and to set local standards. This will involve:-

- defining the services which a local data centre could provide including the cost for providing this service, the benefits to each customer organisation and a model management and funding structure;
- a set of locally and nationally validated data management policies for local data centres;
- guidance on the types of data which will be of most value to Local Biodiversity Action Plans, and the service level required for each customer organisation; and
- the provision of a nationally provided support service for the Recorder Software Package.

This three year stage will cost £520,000.

6.32 The second stage is to support local initiatives and to establish or enhance local data centres. This will involve:-

- a target of 50 local initiatives to establish or enhance local records centres to help meet the long term information needs of Local Biodiversity Action Plans;
- a target of 50 local data centres to adopt standard policies, and ensure that local data contributes to Local Biodiversity Action Plans and is accessible via the data catalogue;
- a number of schemes to improve the value and relevance of voluntary contributions to local biodiversity information which will involve partnerships between statutory and voluntary bodies; and
- facilities to exchange local species information between local areas and nationally.

This stage can be tuned to the pace of Local Biodiversity Action Plan initiatives, and can start as soon as the consortium funding case produced in the first stage is agreed. The stage would cost £1.175m and last at the minimum three years.

6.33 The third stage will be to build local systems and networks. This will involve:-

- a report documenting the progress in establishing consortium funding for local data centres, a review of the viability of this approach, and an assessment of the value of the service provided;
- identifying any new or changed information systems which would improve the service provided by local data centres, and increase the value of local biodiversity data;
- the establishment of network links between local data centres and consortium members; and
- a new or revised information system to improve local biodiversity data management.

Stage three could cost £725,000 over a minimum of four years, and is best started when a number of local consortium funded data centres have been established in stage two.

6.34 Table 10 shows the estimated cost for these three stages. The costings in the table do not include the costs of running local data centres which need to be sought from local formed consortia.

6.35 The proposed information management strategy will involve the co-ordination of activities across a number of organisations. It will involve the change of working practices, and if consortium funding of local data centres is to succeed, the strategy will require changes to the way local organisations budget and pay for data and information management. At a national level, the strategy will influence the established information policies of a number of national organisations.

Table 10: THE LOCAL APPROACH

Project	1996/7 (£000)	1997/8 (£000)	1998/9 (£000)	1999+ (£000)
Stage One				
Devise consortium funding case	60			
Standards for data centre policies	45	25		
Develop guidance on priority datasets	45	45		
Support for Recorder	65	65	65	
Management overhead	35	35	35	
TOTAL	250	170	100	
Stage Two				
Support establishment of consortium funding		60	145	210
Support establishment of standard policies		25	50	50
Support making high priority data accessible		80	160	160
Development of Recorder		100		
Support for IT investment		30		
Management overhead		35	35	35
TOTAL		330	390	455
Stage Three				
Review viability of consortium funding			60	
Review data centre systems needs			60	30
Support for linking data centres				250
Develop systems identified by the review				255
Management cost			35	35
TOTAL			155	570
GRAND TOTAL	250	500	645	1025

PROPOSALS TO INCREASE PUBLIC AWARENESS AND INVOLVEMENT

6.36 To increase public awareness and involvement, we see a need for action in four areas: Government stimulated action, local action, key sectors and education in its broadest sense.

Government Stimulated Action

6.37 The Government has lead responsibility in raising the profile of biodiversity as an issue on the public agenda. There are many ways this could be achieved, but to kick-start the process and give it impetus, we propose a series of high level meetings with senior decision makers from all sectors of society to secure their involvement and to generate ideas. As we explain below, there are particular roles for local authorities, the business community and those involved in education, but all parts of society have a stake in conserving biodiversity, and the initial round of meetings should aim to involve as many special interests as possible.

6.38 The Government should build on this momentum, and provide information to the public by way of an Annual Statement on the Environment, in which progress on biodiversity conservation would form a major element.

6.39 Government departments and agencies can also ensure that biodiversity is given proper consideration in all relevant policy issues, for example in seeing that Circulars, PPG Notes and other guidance dealing with planning and the use of the natural environment are updated to take account of the UK Biodiversity Action Plan.

6.40 We propose that the conservation of biodiversity be incorporated as a criterion for determining appropriate Government (and local authority) grants and awards. New schemes of grant-aid can be promoted, or existing ones adapted, using the success of the Darwin Initiative as a model. Research councils should also be encouraged to place a high priority on research of direct relevance to biodiversity.

6.41 The Government also has a crucial role to play in providing information which helps individuals to make connections between their day to day behaviour and its effect on biodiversity. Examples would be the effects of various modes of transport, or the opportunities for encouraging nature around the home. As a high profile start, the Government may wish to consider introducing an Annual Biodiversity Day, and to re-introduce National Nature Week (the equivalent of National Tree Week, which has proved successful).

Local Action

6.42 There are a number of strategic elements which we see as crucial to action at a local level. Many of these measures fall to local authorities to initiate, but there are others where central Government has a part to play - either directly, or by creating suitable conditions for initiatives to be started or pursued. We also envisage that the Wildlife Trusts and other local and national voluntary bodies, together with business and industry, will have a vital role in prompting and supporting local projects.

6.43 The foremost requirement is the need to raise awareness of biodiversity within Local Government, and to encourage local authorities to commit themselves to action. Local authorities not only have responsibility for carrying out important local conservation work and funding local activities, but through their regulatory duties they can control the actions of others. Local authorities should therefore consider fully the implications for biodiversity when taking planning decisions and carrying out public works, and should encourage others to have regard to biodiversity issues.

L Gill

Buxbaumia Viridis

6.44 At grass root level, local action requires the participation of the inhabitants of any locality, whether it is a hamlet of a few houses in a remote part of the country, or a county or region. Local authorities have an important role to play in raising awareness of biodiversity within local communities, and encouraging opportunities for action - ie competing for central or European funds for individual initiatives. Whatever the level of local action, there are a number of prerequisites which we see as desirable, even essential, in raising public interest and enthusiasm. Some examples are:-

- promoting neighbourhood nature areas where local people can experience and enjoy nature at first hand;
- promoting Local Ecology Centres to provide a focus for public involvement and education, and providing the impetus to develop networks within the local community;
- establishing a network of records centres including new centres where none currently exists; and

- promoting local action to protect biodiversity as part of other initiatives on sustainable development.

6.45 These community projects will provide a foundation for developing Local Biodiversity Action Plans and publicising information about local wildlife, and will give opportunities for local people to become involved in recording schemes.

Key Sectors

6.46 We believe that the most appropriate way to raise awareness of biodiversity is to disseminate information in accordance with the particular interests and concerns of each key sector. To begin this process, we recommend that "champions" should be identified who would have the respect and confidence of their colleagues, and who could help to promote key messages in their sector.

Large Blue Butterfly

6.47 To engage the key sectors, gain their commitment and mobilise action, the high level meetings discussed in paragraph 6.40 should take place. Part of the function of these meetings will be to produce specific messages appropriate to each sector which can be disseminated using internal publications and other means.

6.48 The media have an important role to play in raising the profile of biodiversity, and it is crucial to the success of biodiversity conservation that proprietors and editors are encouraged to give greater prominence to biodiversity issues at both national and local levels. The specialist press (horticultural and country sports magazines, ramblers' newsletters, travel magazines for example) have a role in promoting specific biodiversity issues.

6.49 Because of its involvement and potential influence on biodiversity at several levels, the business community is another key sector. Industry and business should be encouraged to take account of the need to conserve biodiversity, and to raise

greater awareness of biodiversity among their staff. Local businesses could become involved in the production of Local Biodiversity Action Plans, as potential sponsors. It would be helpful for environmental audit of business activities to address implications for biodiversity. Trade Unions and professional associations could also be encouraged to promote awareness of nature as a means of improving the working environment.

Education and Training

6.50 The role of education and training is to promote increased knowledge and understanding of biodiversity as part of a UK Strategy for Environmental Education. This will involve raising the profile of biodiversity within the formal education sector and:-

- ensuring sufficient emphasis is given in vocational and professional training schemes on the principles and practice of biodiversity;
- raising awareness among professions closely involved with biodiversity management;
- providing training to meet the specific needs of the Biodiversity Action Plan; and
- encouraging greater public involvement with nature as a source of education and enjoyment. The Scottish Office report *Learning for Life* is commended as a model for environmental education and as a basis for policy development.

6.51 In England, the Department of the Environment, with the support of the Department for Education and Employment, has commissioned Professor Toyne to conduct a survey of the extent to which the recommendations contained in his Committee's Report on Further and Higher Education have been implemented. The review will identify and promulgate good practice in integrating the environment into curriculum provisions and institutional management, and review practice in promoting mutual updating of environmental information between academia and business.

6.52 We have identified a number of elements essential to improving education, and the teaching of biodiversity and environmental matters generally. The first requirement is to ensure adequate training for teachers, in particular for primary teachers, through initial training at college or through in-service training at schools. The second element is to ensure that biodiversity conservation is one of the topics included under requirements for assessment and review. Knowledge and understanding of biodiversity should also be incorporated into relevant National Scottish Vocational qualifications.

6.53 In the working world, relevant professional bodies should consider whether their institutional qualifications should incorporate knowledge and understanding of the principles and practice of biodiversity conservation. They should also promote biodiversity training as part of continuing professional development. Coupled with this, employers should make a commitment, where appropriate, to training in biodiversity conservation as part of continuing vocational development.

6.54 In the conservation world, training courses should be arranged for biodiversity practitioners in both public and voluntary sectors to increase understanding of the overall approach and, in particular, awareness of the national and local targets. It would be useful to provide training for participants in the Local Action Plan process, in the same way that community participation training is being developed for Local Agenda 21. Field experience of nature should be integrated in formal education and encouraged as an informal activity in the local community.

Timing

6.55 We consider that the programme for Government stimulated action, local action and action by key sectors should be implemented over the next three years, and the programme for education and training over the next five to ten years. Implementing these proposals will help to meet the UK's obligation under Article 13 of the Convention on Biological Diversity which requires contracting parties to:-

> *"Promote and encourage understanding of the importance of, and the measures required for, the conservation of biological diversity, as well as its propagation through media, and the inclusion of these topics in educational programmes".*

6.56 A more detailed list of proposals is shown at Annex D.

IMPLEMENTATION

6.57 To achieve an effective outcome, there is a need to oversee the implementation of the proposals contained in this report. Some tasks may be handled by a UK Biodiversity Secretariat, for example providing a general information service and organising workshops and seminars. But many of our proposals will require a Focus Group, or Groups, to kick start and oversee the necessary action. This includes:-

- finalising work on the long and middle species lists and their monitoring status;
- implementing individual species and habitat action plans;
- completing the preparation of outstanding plans;
- advising on good practice and consistency with the preparation of Local Biodiversity Action Plans;
- taking forward work on information and data, in particular the co-operative network to be overseen by the JNCC and

the development of local consortiums; and
- promoting public awareness and involvement through Government stimulated action, local action, promoting programmes of action by key sectors and the role of environmental education.

6.58 There is a strong case for having Focus Groups at country level to implement the relevant programmes in their countries, since institutions in both the private and public sectors tend to be organised at country level. However, some issues transcend national boundaries, for example information and data and public awareness and involvement. In addition, responses to international commitments need to be handled at the UK level. Many of our proposals are for the UK as a whole, although some require implementation at the regional or local level.

Lowland Heath

6.59 We therefore propose the establishment of a National Focus Group, supported by Country Focus Groups as appropriate. To keep bureaucracy to a minimum, the membership of the respective Groups should be restricted to a small number of organisations with primary responsibility for implementing the proposals in this report. The National Focus Group should be supported by a small UK Biodiversity Secretariat (see below), and the Country Focus Groups by small Secretariats provided by the respective Environment Departments.

6.60 We invite the Government to follow this approach, and to agree a remit for the National Focus Group and the Country Focus Groups.

Implementation of the Targets

6.61 Responsibility for prioritising and delivery of the species and habitat plans will fall to the Government and its statutory nature conservation agencies. However, to stimulate action and to produce commitment we suggest that a 'champion' should be invited to co-ordinate each plan and to act as a facilitator and catalyst by establishing an information network for the plan and generally stimulating the appropriate action. Several organisations have already expressed interest in this, and we invite the Government to endorse this approach, and to agree who should act as the champion for each plan.

6.62 For habitat plans, we would normally expect a statutory organisation to act as champion. For species plans, the champion could be either a statutory or non-statutory organisation. Some organisations are closely identified with species, for example the RSPB with the corncrake, the Forestry Commission with the red squirrel, and the Game Conservancy with the grey partridge. They may wish to take on the co-ordinating role being proposed. There is also the opportunity to invite the business sector to sponsor individual plans, and some companies have expressed an interest in this role.

Lagoons

UK Biodiversity Secretariat

6.63 We recommend that a small, but full time unit, of perhaps two or three people be established as the UK Biodiversity Secretariat. This unit would provide advice within Government, co-ordinate follow up work, maintain the link with the international convention, service the proposed National Focus Group and organise workshops and seminars. The unit should be located in DOE.

6.64 The advantage of such a unit would be in helping to provide continuity and expertise in this area, and the resources required to drive work forward so that proposals do not "wither on the vine". Where there is specific follow up work, for example completing the preparation of outstanding habitat and species action plans, or with Local Biodiversity Action Plans, consideration could be given to secondees to the Secretariat from agencies, local Government or conservation bodies as appropriate.

MONITORING AND REVIEW

6.65 The Steering Group attaches high importance to monitoring key species and habitats in a cost effective way. This will require close co-operation between the voluntary and statutory sectors to maximise the efforts of both. The work should be driven forward by the Focus Groups with the support of the Secretariat and the statutory agencies.

6.66 Monitoring should begin immediately for the short and medium species lists described in Chapter 2, and the list of key habitats. Monitoring the long list of species, (see Annex F), is also of importance in establishing a review of the health of biodiversity in the UK. Although the extent of the work required to produce baseline information is not yet complete, the aim should be to monitor these species, at least in so far as to detect undesirable change, within five years.

6.67 The monitoring programme will need to be modulated to take account of European and international obligations and existing survey commitments.

Reporting Progress

6.68 General progress is picked up by the Department of the Environment's Annual Report on the state of the environment (the annual White Paper). Input to this could be co-ordinated by the proposed UK Biodiversity Secretariat. We also propose that the Secretariat should have responsibility for co-ordinating the UK national report to the Biodiversity Convention.

6.69 In addition, we see a need for a regular report on progress in delivering the targets, and the implications which arise from monitoring the species and habitat action plans. We suggest that this might be undertaken every five years by the National Focus Group.

CONCLUSION

6.70 This report is a positive attempt to meet the UK's international obligations and to promote the conservation of biodiversity at the national and local level. We commend it to all concerned - Government, local authorities, non-Governmental organisations and other sectors alike.

ANNEXES

ANNEX A

THE REMIT AND COMPOSITION OF THE STEERING GROUP

INTRODUCTION

1 As proposed in *Biodiversity:The UK Action Plan*, a Biodiversity Steering Group was established with members drawn from different sectors. This helped to sustain the open process by which the plan was produced and strengthened collaboration.

2 Individuals were invited to become members of the Group on the basis of their expertise and potential contribution to the process as a whole. The Steering Group Report therefore reflects, collectively, individual's advice and not the views of Government departments, agencies, academia and other organisations.

3 The Group, which held its first meeting in June 1994, had as its immediate remit overseeing:-
- the development of a range of specific costed targets for key species and habitats for the years 2000 and 2010 to be published in European Nature Conservation Year 1995;
- improving the accessibility and co-ordination of existing biological datasets, to provide common standards for future recording and to examine the feasibility in due course of a single UK Biota Database;
- the preparation and implementation of a campaign to increase public awareness of, and involvement in, conserving UK biodiversity;

- the establishment of a review process for the delivery of the commitments contained in the plan.

4 In addressing these issues, separate Sub-Groups on Targets, Data, Public Awareness and Local Biodiversity Action Plans were set up. The membership of the main Steering Group and Sub-Groups are shown in the attached box.

5 The Group would like to acknowledge the strong support it has received from many individuals including those from the voluntary sector. The UK is fortunate in having a long tradition of survey and research in natural history, and this tradition and interest is reflected in the strength of membership of those voluntary organisations which are concerned about the conservation of wildlife.

6 A group of these bodies consisting of Butterfly Conservation, Friends of the Earth, Plantlife, The Wildlife Trusts, The Royal Society for the Protection of Birds and Worldwide Fund for Nature joined together to form *Biodiversity Challenge*. *Biodiversity Challenge* has published an agenda for conservation action in the UK, and has made an active and positive contribution to the work of the Group.

BIODIVERSITY ACTION PLAN STEERING GROUP - MEMBERSHIP

Departments

John Plowman (Chairman)	Department of the Environment
Roy Bunce	Department of the Environment
Stephen Hampson/John Randall	Scottish Office
Alwyn Jones/Derek Beames	Welsh Office
John Faulkner	Department of Environment (Northern Ireland)
John Robbs/Peter Boyling/Judy Allfrey	Ministry of Agriculture, Fisheries and Food
Frances MacLeod	Foreign and Commonwealth Office
Phil Ratcliffe	Forestry Commission

Agencies

Derek Langslow	English Nature
Michael Usher	Scottish Natural Heritage
Malcolm Smith	Countryside Council for Wales
Roy Walker	Joint Nature Conservation Committee
Roger Clarke/Penny Jones	Countryside Commission
John Krebs/Mike Roberts	Natural Environment Research Council
Paul Raven	National Rivers Authority

Collections

Ian Gauld/Ian Tittley/Emma Watson	Natural History Museum
John Simmons/Andrew Jackson	Royal Botanic Gardens, Kew
David Rae/Helen Jones	Royal Botanic Garden, Edinburgh

Local Government

Steve Ankers	East Sussex County Council
David Goode	London Ecology Unit
John Sheldon	Lothian Regional Council/COSLA

- MEMBERSHIP continued

Academia

Martin Angel	Institute of Oceanographic Studies
Charles Gimingham	University of Aberdeen
Chris Perrins	University of Oxford, Department of Zoology

Industry

Charlotte Grezo	International Petroleum Industry Environmental Conservation Association
Guy Masdin	Research Management Consultant

Farming and Land Management

Oliver Doubleday	National Farmers Union
Alan Woods	Country Landowners Association

Voluntary Conservation Organisations

David Erwin	Ulster Wildlife Trust
Susan Gubbay	Wildlife and Countryside Link Marine Group
Robin Pellew/Chris Tydeman	Worldwide Fund for Nature (UK)
Tim Sands	The Wildlife Trusts
Graham Wynne	Royal Society for the Protection of Birds

Secretariat

Roger Bendall	Department of the Environment
John Robbins	Department of the Environment
Martin Steer	Department of the Environment
Jo Harrison	Department of the Environment
Madeline Fox	Department of the Environment

TARGETS SUB GROUP

Derek Langslow (Chairman)	English Nature
Martin Angel	Institute of Oceanographic Studies
Leo Batten	English Nature
Roger Bendall/Sarah Webster/ John Robbins/John Corkindale	Department of the Environment
Roger Clarke	Countryside Commission
Charlotte Grezo	International Petroleum Industry Environmental Conservation Association
Maurice Gosling	Institute of Zoology
Susan Gubbay	Wildlife and Countryside Link Marine Group
Sandy Kerr	Scottish Natural Heritage
Chris Perrins	University of Oxford
Phil Ratcliffe	Forestry Commission
Paul Raven	National Rivers Authority
John Robbs/Peter Boyling	MAFF
Malcolm Smith	Countryside Council for Wales
Ian Tittley	Natural History Museum
Roy Walker	Joint Nature Conservation Committee
Graham Wynne	RSPB

Secretariat and Editorial Team

Mairi Cooper	Joint Nature Conservation Committee
Sue Davies	Joint Nature Conservation Committee
Martin Steer	Department of the Environment
John Taylor	RSPB
Rob Wolton	English Nature

DATA SUB-GROUP

Roy Walker (Chairman)	Joint Nature Conservation Committee
Martin Angel	Institute of Oceanographic Studies
Mark Avery	RSPB
Leo Batten/Mark Felton	English Nature
Roger Bendall	Department of the Environment
Sir John Burnett	Co-ordinating Commission for Biological Recording
John Busby	World Conservation Monitoring Centre
Bill Ely/Nicky Court	National Federation of Biological Recording
David Erwin	Ulster Wildlife Trust
Jeremy Greenwood	British Trust for Ornithology
Paul Harding	Institute of Terrestrial Ecology
Sara Hawkeswell	The Wildlife Trusts
Ed Mackey	Scottish Natural Heritage
David Rae	Royal Botanic Gardens, Edinburgh
Tim Reed	Joint Nature Conservation Committee
Mike Roberts/Barry Wyatt	Institute of Terrestrial Ecology
Terry Rowell	Countryside Council for Wales
Lawrence Way	Joint Nature Conservation Committee
Sarah Webster	Department of the Environment
Alan Woods	Country Landowners Association
Gy Ovenden (Secretary)	Department of the Environment

PUBLIC AWARENESS AND INVOLVEMENT SUB-GROUP

David Goode (Chairman)	London Ecology Unit
Steve Ankers	East Sussex County Council
Gillian Beauchamp/Clive Griffiths	Department for Education
Roger Bendall	Department of the Environment
Ian Dair	English Nature
Charles Gimingham	University of Aberdeen
Rosemary Griggs	Department of the Environment
Sharon Gunn	English Nature
Carolyn Harrison	University College London
Doug Hulyer	Wildfowl and Wetlands Trust
Andy Kirby	Department of the Environment
Hilary Le Fort	Department of the Environment
Andy Macpherson	Department of the Environment
Tim Sands	The Wildlife Trusts
John Sheldon	Lothian Regional Council/COSLA
Monica Straughan	Scottish Natural Heritage
Sally York	Forestry Commission
John Robbins (Secretary)	Department of the Environment

LOCAL BIODIVERSITY ACTION SUB-GROUP

David Goode (Chairman)	London Ecology Unit
Steve Ankers	East Sussex County Council
Leo Batten	English Nature
Roger Bendall	Department of the Environment
Bill Butcher	The Wildlife Trusts
Jaqui Cuff	RSPB
Sara Hawkeswell	The Wildlife Trusts
Mike Oxford	Avon County Council
Tim Sands	The Wildlife Trusts
John Sheldon	Lothian Regional Council/COSLA
John Taylor	RSPB
Tony Whitbread	The Wildlife Trusts
Alan Woods	Country Landowners Association
John Robbins (Secretary)	Department of the Environment

DEPENDENT TERRITORIES PROGRESS REPORT

INTRODUCTION

I To date, the British Virgin Islands, the Cayman Islands, Gibraltar, Jersey and St Helena and its Dependencies have been included in the UK's ratification of the Biodiversity Convention. Their Governments have taken a number of steps to protect the biodiversity in their care.

British Virgin Islands

2 The BVI Government is moving towards comprehensive environmental legislation. In August the Chief Minister's office sponsored a meeting to address institutional building, environmental legislation, methodologies and integrated planning issues for sustainable development. It was agreed that there is a need to move towards formulating effective environmental legislation for the sustainable development of the BVI.

Cayman Islands

3 Approximately 4.7% of the land area of the three islands is now protected. This includes several properties owned by the Cayman Islands' National Trust.

4 An extensive system of marine protected areas has been in place since 1986. Programmes to monitor components of marine biodiversity include a comprehensive coral reef monitoring programme and annual assessment of both the status of the adult shallow water queen conch (strombas gigus) populations and the artisinal Nassau grouper (Epinephelus striatus) fishery. Mangrove and seagrass monitoring sites have also been established.

5 A number of the saline coastal ponds are protected as Animal Sanctuaries under the Animals Law (1976) and Regulations: Colliers Pond and Meagre Bay Pond on Grand Cayman, Salt-water Pond on Cayman Brac and the Booby Pond Little Cayman. The Booby Pond (home to one of the largest breeding colonies of red-footed boobies (sula sula) in the western hemisphere) has also been designated as a wetland of international importance under the Ramsar Convention. In addition, the Environmental Zone of the Marine Parks (which include 1600 acres of the environmentally critical interface between Grand Cayman's Central Mangrove Wetland and Little Sound) plus an additional parcel of Crown Land has recently been designated as an Animal Sanctuary. Following public consultation (which is currently underway) the area will also be designated as a Ramsar site. This is intended as an initial step towards securing further protection of the majority of this economically and ecologically important wetland system.

6 Single species conservation programmes taking place in the terrestrial environment include the National Trust's captive breeding programme for the endangered Grand Cayman Blue iguana and a planned ex situ conservation programme for the threatened endemic bromelid (Hohenbergia caymanensis). The Trust is also currently funding a study of the Grand Cayman population of the West Indian Whistling duck to determine their habitat requirements and nutritional needs.

Gibraltar

7 Gibraltar is ecologically significant and sensitive. Its significance stems from its location on a major migration route, but it also has a very substantial flora for so small an area as well as important marine biological assets. Its sensitivity is in part because of its small area, with a high human population density, together with its vulnerability to environmental damage in neighbouring sea and land areas.

8 Much sound scientifically-based work has been carried out in Gibraltar over a number of years. It has a good, but far from comprehensive database. It has good legislation in place to protect the environment, most notably the Nature Protection Ordinance introduced in 1991. A large area of the Upper Rock is designated as a Nature Reserve. It is well protected and managed. In addition, as a territory in the European Union, Gibraltar is required to transpose relevant EU Directives. These include the Habitats Directive which Gibraltar has already implemented.

9 Gibraltar is also fortunate in having excellent human resources to back up the conservation efforts. A partnership exists between the Government and the principal Non-Government Organisation, the Gibraltar Ornithological and Natural History Society who provide expert advice on a wide range of environmental issues and has prepared a report on maintaining the biodiversity of Gibraltar.

10 Gibraltar is being considered as the location for one of the global Geographical Observatories under the programme initiative by the Royal Geographical Society. The main problem which Gibraltar faces, in common with many others, is lack of funds. Gibraltar has made strenuous efforts to obtain external funding for its biodiversity work, with limited success.

Jersey

11 A draft biodiversity strategy has been prepared for Jersey. This adopts an approach based on targets for priority species and ecosystems. These have been selected with respect to their international, British Isles and Jersey importance and the fact that their care and protection will ensure the conservation of biodiversity in Jersey. The strategy includes recommendations for the key issues affecting biodiversity in Jersey, such as policy and planning, agriculture and water supply. Certain management and monitoring tasks are already underway and a full-time Interpretation Officer has been appointed.

St Helena

12 A local review of environmental legislation has taken place and a new ordinance is being drafted to provide for the protection of endangered, endemic and indigenous species. The St Helena Government have accepted the recommendation to adopt a Sustainable Environment and Development Strategy (SEDS) for St Helena and an Advisory Committee on the Environment (ACE) is being established to integrate conservation into national decision making.

13 In addition, the St Helena Government has undertaken the following activities:

- a habitat conservation plan has been drafted for the Peaks area. New photo and ground monitoring plots have been set up to assess the long term impact of habitat management activities in the area and conservation work has started to control the invasive weeds;
- species recovery programmes have been drafted for critically endangered species of endemic plants. The Agriculture and Fisheries Departments' Environmental Conservation Section has increased propagation of key endemic species in order to establish seed orchards and supply stock for habitat restoration work;
- an arboretum and nature trail with on site interpretation facilities has been set up;
- a number of experts have visited St Helena to advise on local plant conservation and to study the local flora subterranean invertebrates and avian fauna and subterranean aquatic fauna;
- the Government has liaised with the fledgling St Helena Conservation Group and the Heritage Society to encourage their development as conservation NGOs;
- an 18 month post has been created to focus on endemic species conservation through habitat restoration and environmental education;
- regular articles are published in the local media to raise awareness of biodiversity issues on the Island. A one day seminar on Sustainable Ecology was held in March.

ACTION IN TERRITORIES NOT YET INCLUDED IN THE UK'S RATIFICATION OF THE CONVENTION

Hong Kong

14 Two biodiversity surveys will shortly be carried out in Hong Kong: one on terrestrial and freshwater habitats, the other on corals and fishes. The surveys will provide an up to date database needed for developing a strategy for biodiversity conservation in Hong Kong. In addition the Mai Po Marshes and Deep Inner Bay has recently been listed as a wetland of international importance under the Ramsar Convention.

Falkland Islands

15 The Attorney General in the Falkland Islands is giving priority to a revision of environmental legislation in 1995. The new legislation should be in place by 1997. Several baseline surveys of the biodiversity in the Falkland Islands are underway.

Montserrat

16 A number of biodiversity conservation activities have been undertaken on the island:

- all lands have been zoned to facilitate development and conservation on the basis of land capability;
- a survey of the bat, frog and lizard population has been completed. This was done jointly by the Montserrat National Trust, Flora and Fauna International and the Forestry Division of the Ministry of Agriculture, Trade and the Environment;
- draft Forestry, Wildlife and National Parks Legislation is now before the Executive Council and should be enacted shortly;
- a number of reforestation efforts have been undertaken primarily for watershed management, soil and water conservation and for the creation of suitable wildlife habitat;
- a revised Animals and Trespass and Pounds Ordinance is being vigorously enforced to arrest the adverse effects of feral livestock on the terrestrial and marine environments;
- a baseline survey and management plan was conducted for the Foxes Bay Wildlife Reserve. This was a partnership effort of the University of the West Indies, the Montserrat National Trust and the Forestry Division. Montserrat is actively considering the necessary steps to enable it to sign the Biodiversity Convention.

Pitcairn Islands

17 A draft management plan has been produced for the environment on Henderson Island in the Pitcairn Islands. This is currently with the Island Council for approval.

British Indian Ocean Territory

18 The University of Warwick is leading an expedition to the British Indian Ocean Territory in early 1996. This will draw up a conservation plan for the Territory.

Guernsey

19 A baseline study of Guernsey flora and fauna is in the final stages of completion, and this will be followed by consideration of a strategy for the implementation of such protection measures as are considered necessary.

Isle of Man

20 The Isle of Man has enabling legislation in place. Its Department of Agriculture, Fisheries and Forestry has conducted a baseline survey of the Island to help in determining which areas are worthy of specific protective measures. It is now moving onto a second phase survey which will look at the interesting areas in more detail with an eventual view to deciding which should be designated as areas of specific interest.

ANNEX C

GUIDANCE IN DEVELOPING LOCAL BIODIVERSITY ACTION PLANS

INTRODUCTION

1 In June 1992, the Prime Minister and over 150 Heads of State or Governments, signed the Convention on Biological Diversity at the "Earth Summit" in Rio de Janeiro. Signatories recognised that action must be taken to halt the world-wide loss of animals and plant species and genetic resources. They recognised that each country has the primary responsibility to conserve and enhance biodiversity within its own jurisdiction. At the same time, they agreed to draw up national plans and programmes and to share resources to help implement such programmes. The Convention on Biological Diversity is essentially a commitment to conserving and sustaining the variety of life on earth.

2 At the launch of *"Biodiversity: The UK Action Plan"*, in January 1994, the Prime Minister announced that a national Biodiversity Steering Group would be established to develop work in this field. This Steering Group has addressed four main areas, namely:-

- UK targets for habitats and species;
- data requirements;
- raising public awareness and involvement; and
- guidance for the production of Local Biodiversity Action Plans.

3 The Steering Group reported in December 1995 with detailed proposals on each of these topics as the basis for implementation of the UK Action Plan. It was agreed that the guidance for production of Local Biodiversity Action Plans should be made widely available as a separate document for all those likely to be involved in this process.

4 If the UK Biodiversity Action Plan is to be implemented successfully it requires some means of ensuring that the national strategy is translated into effective action at the local level. Local Biodiversity Action Plans are seen as a means by which such action can be achieved. One of their main functions is to ensure that national targets for species and habitats are attained in a consistent manner throughout the UK. But there is a great deal more to such plans than simply providing a mechanism for meeting the national targets. They provide a means for delivery of several objectives of the UK Action Plan.

5 One of these objectives is to promote the conservation of species and habitats characteristic of local areas. Local Biodiversity Action Plans provide the focus for local initiatives to fulfil local needs in terms of biodiversity conservation. They provide an opportunity for local people to express their views on what is important. Local plans should seek, therefore, to include targets which reflect the values of local people and

which are based on the range of local conditions, and thereby catering for local distinctiveness. However, since it will largely involve land which is in private ownership, the approach will require considerable consultation, guidance and involvement of various sectors, to create the new working partnerships necessary for success.

6 It is expected that a Local Biodiversity Action Plan will act as a catalyst to develop effective partnerships capable of ensuring that programmes for conservation of biodiversity are maintained in the long-term. These local partnerships will, in turn, assist in raising awareness of the importance of biodiversity, thereby gaining wider public commitment. Joint "ownership" of a Local Plan is regarded as crucial to success in building commitment within the local community. Production of a Local Biodiversity Action Plan will provide the biodiversity component of Local Agenda 21.

7 Local Biodiversity Action Plans should also identify where it is appropriate and cost-effective to halt recent trends in habitat fragmentation, and create new and attractive landscapes by habitat enhancement and restoration.

8 Such plans will have a key role in monitoring progress in the conservation of biodiversity in the long-term. The data which form an integral part of Local Biodiversity Action Plans needs to be compatible with the national biodiversity database.

9 In summary:-
The purpose of Local Biodiversity Action Plans is to focus resources to conserve and enhance the biodiversity resource by means of local partnerships, taking account of both national and local priorities.

GENERAL APPROACH

10 A Local Biodiversity Action Plan is both a product and a process. Not only does it identify where action needs to be taken to implement the national targets for habitats and species, but it also specifies appropriate delivery mechanisms. The scale of such Plans can be expected to vary considerably between different parts of the UK, for example they might be developed at local authority level. In England they should ideally be developed at the county or district level of local authorities. In some areas where insufficient data is available, it may be appropriate for Plans to be developed initially at a regional level to ensure that action is co-ordinated and that data is collected to enable further refinement of the Plan to a local level. The process of taking forward Local Biodiversity Action Plans in Scotland should take account of the particular circumstances there.

11 The formulation of Local Biodiversity Action Plans should not be undertaken by a single organisation, but there is a need for one organisation to take the lead. Local authorities are ideally suited to have this role, working with statutory conservation and countryside agencies, local and regional voluntary organisations, land managers, businesses, local records centres and specialist recorders. To be successful, the Plan should be owned by all the parties who have a key role in delivering the product.

12 The process of developing a Plan requires several quite distinct elements. Analysis and evaluation of the nature conservation resource is clearly a major part of the Plan, resulting in detailed proposals for action within a specified period of time. In parallel with this is the development of an effective partnership with key players, particularly land managers, to identify appropriate delivery and funding mechanisms. A third component is the programme for monitoring the effectiveness of the Plan including the extent to which both national and local targets are being achieved. Underlying all of this is the need for an adequate database at the local level, which is also fully integrated with a national biodiversity database.

MAIN ELEMENTS OF A LOCAL BIODIVERSITY ACTION PLAN

- Establishing a Plan Partnership.
- Agreeing broad objectives.
- Reviewing the resource.
- Evaluating the existing resource within the national and local context.
- Developing specific targets and proposals for action.
- Defining areas for action on a proposals map.
- Identifying delivery mechanisms and sources of finance and advice.
- Establishing a long term monitoring programme to measure the effectiveness of the Plan in achieving national and local targets.

ESTABLISHING A PARTNERSHIP

13 The first step is to establish a Plan Partnership, and for a lead organisation to be identified. The local authority is well placed to take the lead in promoting the Plan as part of its Agenda 21 process. Where a local authority is not in a position to take the lead, then it will be necessary to identify another lead organisation. The Plan Partnership will ultimately include all key players with responsibility for the conservation of biodiversity within the plan area, including representatives of land managers. Initial steps should include a period during which land managers are consulted on the objectives, mechanisms and benefits, as their agreement and co-operation is vital. The Partnership should be established at the outset to encourage the development of a shared vision, and to ensure that all participants are committed to the development of the Plan.

EXAMPLES OF KEY PLAYERS IN A PLAN PARTNERSHIP

Local authority (members and officers)	Local Conservation and amenity groups
Land managers, business and industry	BTCV/SCP
Statutory nature conservation agency	Forestry Authority
Local biological records centre	National Trust (or NTS)
Local wildlife trust/RSPB	FWAG: Woodland Trust
DOE/Scottish Office/ Welsh Office/DOE(NI)	Game or wildfowling organisations
NRA/drainage authorities	Marine conservation bodies.

AGREEING BROAD OBJECTIVES

14 There are two main objectives of a Local Biodiversity Action Plan - to reflect and implement national priorities and, at the same time, take into account local considerations. One of the primary objectives is to ensure the effective implementation of national targets for both species and habitats at the local level. National targets will thus inform and guide the content of local action plans so that implementation is firmly linked to national priorities. At the same time, a local action plan must take account of the range of biodiversity within the local area and its importance to local people. Other objectives include the effective development of a long term process for biodiversity conservation and a means of monitoring progress.

REVIEW OF EXISTING INFORMATION

15 An early requirement is to review existing information for both species and habitats. Such data varies in both quality and coverage in different parts of the UK, and part of the process in developing Plans is to ensure that minimum standards are met. Areas suffering from inadequate information will need to include actions to improve the information base as a matter of priority. The review of existing information will be a highly selective process concentrating on both information relevant

to UK targets for habitats and species, and information needed to set local targets, ie relating to habitats and species characteristic of the area covered by the Plan. This could also include information about species which have been lost from the local area with a view to their reintroduction.

16 The production of a biodiversity map for the local area will be an essential step prior to assessing the resource, evaluating it and identifying areas for action. A biodiversity map shows the location of important semi-natural habitats together with selected information on the distribution of critical species. Inventories exist for a variety of habitats including, for instance, peatland, heathland, grassland, saltmarshes, and other comprehensive information is available for many parts of the UK in the form of Phase I survey data. There is considerable variation in the availability of species information as this depends on individual recording schemes. Information is also available on the location of designated sites. Together, this information will provide the basis for a biodiversity map. Where local records centres exist, data is more readily available. Elsewhere data will be held by different organisations, and it will be necessary to collate existing data and to identify gaps.

17 Shortcomings of existing data should not be regarded as a reason to delay the process. The "best available data" should be used at the same time ensuring that improving the data is given high priority as part of the ongoing Plan.

18 The collation of data on biodiversity is best carried out by local records centres. In areas where such centres do not exist, or where the information is currently inadequate, this deficiency will need to be addressed as part of the Plan. The establishment of new partnerships may be crucial to the development of local records centres capable of maintaining a Plan Database. The most effective long term analysis of appropriate data will often be through a Geographical Information System (GIS) which will enable updating to be carried out at the local level. Local data systems need to be compatible with each other throughout the UK, and with the national biodiversity database.

19 The role of local records centres is dealt with more fully in Chapter 3. They are an essential prerequisite to the successful implementation of *Biodiversity: The UK Action Plan* at the local level, and records centres should ensure that the partners are kept informed about available datasets and their uses.

20 Where appropriate, local authorities will wish to use the data underlying local plans, and the subsequent monitoring component, as part of their ongoing "State of the Environment" reports.

EVALUATING THE RESOURCE

21 A significant part of the Local Biodiversity Action Plan will be devoted to those species and habitats for which targets have been set nationally. These targets result from a critical analysis identifying priorities for action as, for instance, in the case of globally threatened species, or those which have suffered a significant decline in numbers within the UK. Information concerning the distribution and condition of the species and habitats within the Plan area will be required in order to develop strategies appropriate to local circumstances. Priorities for action will be influenced by the national targets.

22 One of the objectives of *Biodiversity: The UK Action Plan* is the conservation and, where possible, the enhancement of habitats and species characteristic of local areas. In considering local characteristics, priority should be placed on the historical and natural context. In England, the Natural Areas Programme will assist this process, as will biogeographical zones in Scotland. Other strategic plans may also assist with this process, such as coastal zone management plans and river catchment management plans. Certain species may be selected because they are characteristic of the local area and are familiar to local people.

23 Elements of biodiversity which are considered to be significant in the local context should be identified at an early stage in the development of the Plan, so that priority can be given to relevant species and habitats in the collation and evaluation of existing data.

IMPLEMENTATION OF ACTION PLANS FOR HABITATS AND SPECIES

24 Action plans specific to particular habitats and species are needed to achieve the objectives of the local Plan. In the case of those habitats and species for which targets have been set nationally, this will require interpretation of the national targets and action plans at the local level. For other species or habitats, considered to be of local significance, it will be necessary to produce specific action plans. To maintain the standard approach, the same framework should be used as with the UK habitat and species action plans. In all cases, action plans should identify the organisations responsible for delivering the required action, and specify a target for its implementation. In addition to the specific action plans, Local Biodiversity Action Plans will contain generic action plans of various kinds. Examples include:-
- establishment of a local records centre where none exists;
- action plans which can be applied to a particular suite of habitats, eg all kinds of woodlands;
- a strategy for a prime biodiversity area;
- identifying the sources of funds and developing effective policies for allocation.

DEFINING AREAS FOR PRIORITY ACTION

25 As a result of the evaluation, in both the national and local context, specific areas will be identified for priority action.

Prime Biodiversity Areas

26 Areas where particular concentrations of high priority habitats occur are often referred to as prime biodiversity areas. These are not designations but are areas where action is likely to be most cost effective. Such concentrations offer opportunities for proactive programmes aimed not only at managing those sites, but also increasing the level of biodiversity of intervening land through habitat management and enhancement, or by means of habitat re-creation aimed at restoring the natural character of the local area.

Urban Areas

27 Special attention needs to be given to the development of action plans for urban areas, where most people live, as a means of raising awareness of the need for the conservation of biodiversity, and involving local people in positive action as part of the local action plan. A key aspect is the establishment of workable targets for the amount of accessible urban wild space available which can be managed for wildlife and for people's enjoyment of nature.

The Wider Countryside

28 Local Biodiversity Action Plans will need to cater for a host of initiatives which already occur within the wider countryside, such as restoration of ponds, trees and hedgerow planting, and the conservation of headlands. Such activities contribute considerably to biodiversity conservation, and should be encouraged wherever appropriate. Although implementation will be largely dependent on individual land managers, it will be necessary to ensure that generic action plans are developed and implemented through local partnerships. It will also be necessary to link landscape and conservation features with specific habitat action plans. Particular attention will also need to be given to rare and declining species with scattered populations where these are not adequately catered for within existing designated sites or within prime biodiversity areas.

IDENTIFYING DELIVERY MECHANISMS

29 The process needs to be linked to the opportunities of grant funding through local or national schemes to stimulate the interest and co-operation of the land managers on whom the success of the Plan will often depend. The funding map or inventory will need to be kept up to date, and advisory bodies such as FWAG or the key funding bodies such as MAFF or Forestry Authority should be included as lead players.

30 Relevant planning guidance and management advisory documents will need to be available to all participating organisations, especially the organisations responsible for implementing action. Advice on the availability of these documents should be a function of the local records centre.

31 It should be emphasised that Local Biodiversity Action Plans do not require new categories of site designation. Implementation of Plans can be achieved successfully using the range of existing designations and incentives.

32 The lead organisations should ensure that those developing detailed targets for action work closely with individual land managers to agree objectives and to ensure that relevant management advice is available. They should also promote full use of available incentive schemes, including agri-environment schemes and agreements under Section 39 of the Wildlife and Countryside Act 1981, to ensure that appropriate management is introduced and maintained in the future so that local targets can be achieved.

MONITORING

33 The Plan should be monitored regularly to assess the degree to which it has been implemented, and how far it meets the targets for habitats and species. This should be a continuous process as different habitat and species action plans will need to be reviewed to different timescales depending on the information available when developing the Plan and the population dynamics of the species.

34 Monitoring changes in the status of species or habitats is required as part of the long term strategy for a Local Biodiversity Action Plan. However, monitoring can be costly, and it will be necessary to be selective. There are differences in the kind of information needed for the overall Plan and the data needed for monitoring. Monitoring should involve not only the species of UK significance, to provide information on UK trends, but should also be related to species of local significance or local interest. Consideration also needs to be given to the selection of species for monitoring in relation to their public appeal.

35 The monitoring process will need to be integrated with monitoring the UK Plan to help evaluate the role of the local area in contributing to national targets. Progress should be reported at the five-yearly period to be adopted for the UK Plan.

THE STAGES IN THE DEVELOPMENT OF A LOCAL BIODIVERSITY ACTION PLAN

Identify lead organisation	Local authority will generally lead. If not, another body must be identified
Establish partnership	Identify main players including statutory agencies, voluntary sector, land managers and business
Agree broad objectives	Agree overall content of plan. Consult with land managers on objectives, mechanisms and benefits
Collect relevant data	Collect best available data on habitats and species relevant to national targets and local significance
Obtain additional data	Where gaps exist in existing information, collect relevant data. Where no local data centre exists, initiate action to establish a centre
Produce biodiversity map	Collate data and produce map showing distribution of key areas
Define prime biodiversity areas	Overlay species and habitat data to show local clusters of biodiversity. Identify prime biodiversity areas on map
Agree priorities for action	Agree priorities based on national targets, local and community objectives, key sites, species, and prime biodiversity areas
Strategy for prime biodiversity areas	Lead organisation to co-ordinate, using species, habitats, local history, geographical data and objectives of natural areas
Strategy for urban areas	Develop strategy for urban areas with a hierarchy of sites to include local needs for peoples enjoyment
Strategy for other areas	Identify priorities for action within the wider countryside
Develop local action plans	Lead organisation to consult on and develop local habitat and species action plans with targets for national and/or local priorities
Delivery mechanisms	Lead organisation to identify funding and other resources available, including sources of advice
Monitor progress	Monitor and review the whole process. Report locally and nationally

36 The provision of this guidance is only the first step towards implementation of Local Biodiversity Action Plans. It needs to be seen in the context of a whole new target led approach to conservation stemming from *Biodiversity: The UK Action Plan*. Implementation will require more than just the guidance to ensure success. It will require training, pilot projects, community workshops and other means to ensure a fully professional product. It would also be necessary to consult with national bodies, especially those representing local authorities and land managers to ensure that the whole programme can be carried forward effectively.

THE RELATIONSHIP BETWEEN LOCAL AND NATIONAL TARGETS

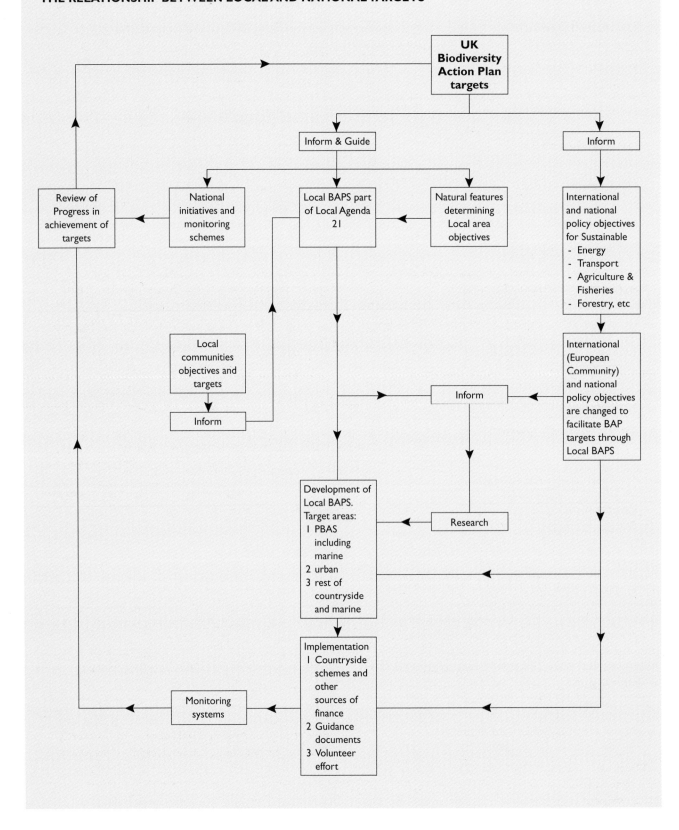

ANNEX D

PROPOSALS TO INCREASE PUBLIC AWARENESS AND INVOLVEMENT

GOVERNMENT STIMULATED ACTION

Provide leadership from the highest levels of Government aimed at raising the profile of biodiversity as an issue on the public agenda.

- Make an Annual Statement on the Environment in which biodiversity is a key element.   *Government*
- Publish an Annual Review of progress on Biodiversity Conservation as part of an Environment White Paper or linked to Annual Reports.   *Environment Departments*
- Set up a series of high-level meetings with senior decision-makers from all sectors of society (in particular the Advisory Committee on Business and the Environment) to kick-start the process and mobilise action in each sector.   *Government*

Raise awareness of the Government's commitment to the Biodiversity Convention in all Departments of Government.

- Provide guidance on biodiversity as part of advice on "green housekeeping".   *Environment Departments*
- Refer fully to biodiversity as a normal part of any statements on progress made in implementing sustainable development. This could be equivalent to consideration of other broad areas of policy such as health and safety.   *Government Departments*
- Ensure that implications of the Biodiversity Convention are considered by individual departments in relation to all relevant policy issues, eg in the next Round of Oil and Gas Exploration; or in development of policy for environmental education.   *Government Departments*

Ensure that biodiversity is given full consideration in arrangements for inter-Departmental collaboration on environmental aspects of policy development.

- Include biodiversity as a topic for consideration by Green Ministers.   *Government*
- Address biodiversity effectively in all Circulars and PPGs dealing with planning or the use of the natural environment; including land-use planning, forestry, fisheries, building materials, extractive industries and waste disposal.   *Government Departments*
- When circulars are revised, ensure that they are updated to take account of the UK Biodiversity Action Plan - especially the setting of targets for action, and the proposals for implementation of Local Biodiversity Action Plans.   *Environment Departments*

Raise public awareness of the importance of biodiversity and its relevance to sustainable development.

- Produce a statement which explains biodiversity as a key component and test of sustainable development.   *Environment Departments*

- Provide information to help individuals to make connections between day-to-day behaviour and its consequent effect on biodiversity; with examples such as the effects of various modes of transport, or the opportunities for encouraging nature around the home.   *Environment Departments*
- Provide guidance on how partnerships might be established to raise awareness of biodiversity and to assist in implementation of the UK Action Plan.   *Environment Departments and Local Government*

Raise the profile of biodiversity as a criterion for grants and awards.

- Promote the importance of conserving biodiversity as a criterion for determining grants and awards.   *Government*
- Promote new schemes of grant-aid, or adapt existing ones, for conserving biodiversity within the UK using the success of the Darwin Initiative as a model.   *Environment Departments*
- Provide a new national award for endeavour in the environment equivalent to the Queens Award to Industry in which biodiversity should play a major part.   *Government*
- Encourage the relevant Research Councils to place a higher priority on research of direct relevance to the conservation of biodiversity.   *Government*

Raise awareness and encourage commitment to implement the UK Biodiversity Action Plan in all sectors of society.

- Communicate the work of the UK Steering Group, by informing the public and specific sectors of progress on all aspects of the Action Plan on a regular basis.   *Environment Departments*
- Issue guidance on development of Local Biodiversity Action Plans.   *Environment Departments*

LOCAL ACTION

Raise awareness and commitment to biodiversity within Local Government.

- Establish a UK Advisory Group to provide leadership and to recommend standards of good practice in the production and implementation of Local Biodiversity Action Plans, and to promote consistency of approach throughout the UK.   *Agenda 21 Steering Group and Local Government Management Board*
- Promote close co-operation with Local Agenda 21 to ensure effective implementation of biodiversity conservation as a crucial part of Agenda 21 through Local Authority Associations and the Local Government Management Board.   *Local Authority Associations*
- Encourage greater awareness of biodiversity conservation amongst local government politicians.   *Local Authorities*

- Issue policy statements committing individual local authorities to action to conserve biodiversity. *Local Authorities*
- Take the lead responsibility in co-ordinating (with other organisations as appropriate) the preparation and implementation of Local Biodiversity Action Plans. *Local Authorities*
- Ensure that biodiversity issues are given proper consideration by all relevant committees of local authorities, including any environmental forum. *Local Authorities*
- Ensure that biodiversity is given full consideration in annual reports reviewing the state of the environment. *Local Authorities*
- Inform the public about local biodiversity conservation as a normal part of local government work. *Local Authorities*
- Ensure that suitably qualified specialists are available, so that local government has the capacity to undertake work on biodiversity in a fully professional manner. *Local Authorities*
- Ensure that continuing professional development includes training in biodiversity conservation for officers in planning and land management. *Local Authorities*
  (See also Training and Education section)

Raise awareness of biodiversity within the local community.

- Establish local networks, or an environment forum, linked to Local Agenda 21, to generate support for local action on biodiversity. *Local Authorities and other appropriate organisations*
- Set up an Action Plan Partnership with the specific purpose of developing and implementing a Local Biodiversity Action Plan. *Local Authorities jointly with other organisations*
- Promote and explain the approach adopted in the UK Action Plan to local groups, networks etc. *Statutory Agencies, Local Authorities and Voluntary bodies*
- Ensure that Guidance is publicly available for development of Local Biodiversity Action Plans, including the reasons for incorporating national targets and the need to meet local objectives. *Statutory Agencies, Local Authorities and voluntary bodies*
- Promote neighbourhood nature areas where local people can experience and enjoy nature at first hand. Such areas are one of the most direct means of raising awareness of biodiversity and involving people directly in conservation action. Accessible local nature areas are equally relevant in rural settlements as in towns and cities. *Local Authorities and The Wildlife Trusts*
- Promote the establishment of Local Ecology Centres to provide a focus for local action and as a venue for public involvement and education. Such centres can provide the crucial impetus to newly developing networks within the local community. *Local Authorities and Environmental and non-Governmental Organisations*

- Develop programmes to raise awareness of biodiversity such as guided walks, nature trails; and opportunities for practical conservation projects through volunteer organisations. *Voluntary bodies*
- Seek to improve standards of interpretation of biodiversity for the general public in parks and local nature areas; and also in school and college grounds. *Statutory Agencies, Local Authorities and Education Authorities*
- Promote local action to protect biodiversity as part of other initiatives on Sustainable Development. *Going for Green and Local Agenda 21 networks*

Raise the profile of biodiversity as a criterion for local grants and awards.

- Promote the importance of conserving biodiversity as a criterion for determining local grants and awards, and introduce a new category of award for local communities based on achievement of biodiversity targets. *Environment Departments and Local Authorities*
- Identify mechanisms for local funding of biodiversity projects. *Statutory Agencies and Local Authorities*
- Introduce a special category of award for local nature areas, Ecology Centres, and school nature gardens. *Local Authorities*
- Establish new awards for "looking after nature" equivalent to the Best Kept Village competition. *Environment Departments*

- Encourage local partnerships to bid for national and European funds for local biodiversity projects. *Local Authorities jointly and non-Governmental Organisations*

Promote public awareness of biodiversity through the establishment of local record centres.

- Support the development of consortium funded local records centres as the best means of supplying the information needed for Local Biodiversity Action Plans and to provide information for the public about local wildlife. *Local Authorities, The Wildlife Trusts etc*
- Promote existing centres and develop the network of record centres to provide a good foundation for Local Biodiversity Action Plans in areas currently without a local records centre focus. *Local Authorities and The Wildlife Trusts*
- Provide opportunities for local people to become involved in recording schemes. *Local Authorities and The Wildlife Trusts*
- Publicise information about local wildlife. *Local Authorities and voluntary bodies*

KEY SECTORS

Raise awareness of biodiversity by dissemination of information advice and guidance through key sectors.

- Identify key sectors for action. *Environment Departments*

- Produce simple statements on what biodiversity is and why it is important. *Environment Departments*
- Produce specific messages appropriate to each key sector. *Environment Departments*
- Identify "champions" who can help to promote key messages in each sector. *Environment Departments*
- Identify key publications, and other means, through which key messages can be publicised. *Environment Departments*

Promote programmes of action in each sector aimed at raising awareness of biodiversity.

- Initiate a series of high-level meetings with key sectors of society to gain their commitment and mobilise action in each sector (see also section on Government Action). *Government*
- Encourage each sector to produce a statement giving a commitment to biodiversity conservation and outlining action to be taken within the sector. *Government*
- Issue a report on behalf of British Industry to give leadership in this field, to explain why biodiversity is important, and to generate action within the sector. *CBI*
- Seek to ensure that biodiversity is given prominence in local partnerships between business and the community aimed at improving the environment. *Local Chambers of Commerce and Agenda 21 networks*
- Promote the involvement of local businesses in the production of Local Biodiversity Action Plans, either as land managers or as potential sponsors. *Business leaders*
- Raise awareness of the need for environmental audit of business activities to address the implications for biodiversity in all aspects of any business. *Environment Departments*
- Encourage greater awareness of biodiversity amongst employers or employees and promote programmes of action for biodiversity conservation within all kinds of businesses (eg rooftop wildlife gardens; or nature reserves within business parks). *CBI*
- Encourage Trade Unions to promote awareness of nature as a means of improving the working environment. *TUC*
- Encourage a high profile for biodiversity issues in all sectors of the media. *Media*
- Encourage proprietors and editors to give greater prominence to biodiversity issues at both national and local levels. *Government*
- Introduce an Annual Biodiversity Day; and re-introduce National Nature Week (as an equivalent of National Tree Week which has been so successful). *Environment Departments*
- Encourage specialist media outlets to promote issues of biodiversity (eg farming, horticultural and gardening magazines, golf, holiday travel, etc). *Media*
- Encourage consideration of topical issues of biodiversity within popular programmes on radio and television. *Media*

EDUCATION AND TRAINING

Promote increased knowledge and understanding of biodiversity as part of a UK Strategy for Environmental Education.

- Promote Learning for Life as a model for environmental education and as a basis for policy development. *Government*
- Recognise the inter-Departmental nature of the process within Government and promote liaison between all Departments and Country Agencies. *Government*
- Pursue a phased approach to the implementation of the Toyne Report on Further and Higher Education setting national and institutional targets, coupled with a monitoring programme. Extend these proposals to Scotland and N Ireland. *Government*
- Identify mechanisms by which partnerships can be established between Government, NGOs and stakeholders at the UK level. *Government and NGOs*
- Promote the establishment of local partnerships including Regional or Local Environmental Education Forums, throughout the UK. *Education Authorities*
- Ensure that local partnerships are closely linked with both Local Agenda 21 (and Going for Green in England and Wales). Build upon the cross-sectoral structures already in place within Local Agenda 21 at the local level. *LA21 Networks and Going for Green*
- Extend the national strategy to include community education initiatives, education in the workplace and continuing professional development. *Government*

Raise the profile of biodiversity within the Formal Education Sector.

- Seek to ensure adequate training in biodiversity for primary teachers, and secondary teachers where relevant, through initial teacher training and through in-service training using local facilities. *Training Providers*
- Review the needs for in-service training in relation to biodiversity competence. *Teacher Training Agency and Institutes of Education in Scotland*
- Encourage the provision of grant aid for training in biodiversity from the GEST programme, from School Effectiveness Grants, and from other LEA sources. *Education Departments and Local Education Authorities*
- Encourage action to identify appropriate material specifically on biodiversity as part of the Government's new initiatives in relation to source material and examples of good practice in environmental education. *Education Departments; SCAA; CEE; and RSPB and other voluntary bodies*
- Publish handbooks on accessible nature areas and the facilities provided within local areas. *LEA*
- Ensure that biodiversity conservation is one of the topics

included for assessment and review.   *HMI Scotland OFSTED Inspectorate*

Raise awareness of the principles and practice of biodiversity within the Professions

- Relevant Professional Institutes should review their requirements for professional qualifications to incorporate knowledge and understanding of the principles and practice of biodiversity conservation.   *Professional Institutes*
- Relevant Professional Institutes should promote biodiversity training as a normal part of continuing professional development.   *Professional Institutes*
- Employers should make a commitment, where appropriate, to training in biodiversity conservation as part of continuing professional development.   *CBI*
- Incorporate knowledge and understanding of biodiversity into relevant National/Scottish Vocational Qualifications.   *COSQUEC*

Ensure that sufficient emphasis is given to biodiversity in training schemes.

- Statutory Agencies and other organisations which fund training should stipulate the biodiversity content of courses as a condition of grant aid.   *Statutory Agencies*
- Provide training courses for practitioners in both public and voluntary sectors to increase understanding of the overall approach and, in particular, to increase awareness of the national and local targets.   *Training Providers; FSC; Statutory Agencies and voluntary organisations*
- Provide training for civil servants and local government officers to raise awareness of biodiversity issues.   *Training providers*
- Provide training for participants in the Local Action Plan process - akin to community participation training being developed for Local Agenda 21.   *Training facilitators*

Encourage greater public involvement with nature as a source of both education and enjoyment

- Promote the direct involvement of people with local wildlife projects as a source of inspiration, learning and enjoyment.   *The Wildlife Trusts; Adult Education; BCTV*
- Encourage field investigations of nature as part of both formal education and as informal activities within the local community.   *FSC; EAs: The Wildlife Trusts; RSPB*

GLOSSARY

Agenda 21
Action plan for the next century, endorsed at the Earth Summit.

Agri-environment schemes
Schemes offering payments to farmers to promote farming compatible with the requirements of the protection of the environment and the maintenance of the countryside under EC Regulation 2078/92. Schemes applicable in all countries of the UK are Environmentally Sensitive Areas, Countryside Access Scheme, Organic Aid Scheme, Habitat Scheme and the Moorland Scheme. Nitrate Sensitive Areas and Countryside Stewardship are applicable in England only and Tir Cymen in Wales only.

Agrochemicals
Chemical substances used in agricultural production including fertilisers, herbicides, fungicides and insecticides.

Ancient woodland
Woodland that is known to have existed before 1600. In Scotland ancient woods are those which were present at the 1750 survey of General Roy.

Arable area payments scheme
The EC system of arable support which offers payments per hectare on the main cereal, oilseed and protein crops. All but the smallest farmers have to set land aside to receive the payments.

Arthropod
Animal with a hard outer casing (exoskeleton) segmented body and jointed limbs, eg an insect or crustacean.

Bern Convention on the Conservation of European Wildlife and Natural Habitats
This imposes obligations to conserve wild plants, birds and other animals, with particular emphasis on endangered and vulnerable species and their habitats. The provisions of the Convention underlie the EC Habitats Directive as well as the UK's wildlife legislation.

Bioaccumulation
The accumulation of substances in living organisms. Where the substances are toxic, this can lead to progressive and irreversible harmful effects.

Biodiversity
The variety of life on Earth or any given part of it.

Biodiversity Convention
The Convention on Biological Diversity. This Convention was signed by the Prime Minister and 150 other Heads of State or Governments at the Earth Summit in Rio de Janeiro in June 1992.

Biogeographical Zones
Major parts of the Earth's surface, including its seas and oceans, characterised by distinctive assemblages of plants and animals.

Biota
All the plants and animals of a given area or region. Thus a UK Biota Database would hold all types of information about the range of species occurring in the UK.

Biotechnology
The application of biological organisms or their products in industrial and chemical processes.

Bonn Convention on the Conservation of Migratory Species of Wild Animals
This requires the protection of listed endangered migratory species, and encourages separate international agreements covering these and other threatened species.

Bryophytes
A major group of plants comprising the mosses and liverworts.

Common Agricultural Policy (CAP)
A European Community wide policy, explicitly provided for in the founding treaty of the Community, which supports agriculture through commodity support measures and market management and through measures to improve agricultural structures.

Common Fisheries Policy (CFP)
A European Community wide policy for fisheries and agriculture which provides for rational and responsible exploitation of living marine resources on a sustainable basis in appropriate economic and social conditions taking account of implications for the marine ecosystem and the needs of producers and consumers.

Contaminant
A substance which is present in elevated concentrations, but not at levels which cause harm or exceed an environmental quality standard.

Convention
An international agreement through which nations agree to work together co-operatively to implement certain defined policies or take other action. International Conventions are voluntarily entered into by countries, although once a country has signed a Convention, it agrees to implement or be bound by the condition specified, eg through its signature of the Ramsar Convention on Wetlands, the UK has agreed to promote the conservation and wise use of all wetlands in the UK.

Convention on Climate Change
Signed by the Prime Minister and other Heads of State and

Governments at the Earth Summit in Rio de Janeiro in June 1992. This Convention commits countries to prepare national programmes to contain greenhouse gas emissions and to return emissions of carbon dioxide and other greenhouse gases to 1990 levels by the year 2000.

Convention on International Trade in Endangered Species (CITES)

This prohibits or regulates international trade in species which are threatened with extinction or likely to become so and are subject to significant trade.

Coppicing

The traditional form of management of much of the broadleaved woodland in the UK, by cutting down trees and shrubs near ground level, allowing the tree to regrow from the stump, and re-cutting at intervals of one or more decades to provide long straight poles.

Countryside Stewardship

An agri-environment scheme which enables farmers and land managers to enter 10 year management agreements to maintain or enhance certain landscapes and features including chalk and limestone grassland; lowland heath; waterside land; coast; uplands; historic landscapes; old traditional orchards; old meadows and pastures; community forests; the countryside around towns and throughout England hedgerows and field boundaries which need restoring. The scheme which was set up and administered by the Countryside Commission completes its five year pilot phase in April 1996 when administration will transfer to MAFF.

Countryside Survey 1990

A survey initiated by the Institute of Terrestrial Ecology and developed by a consortium of other bodies co-ordinated by DOE. Consists of a series of maps covering a sample of 508 1km squares recording details of land cover, field boundaries of habitats and soils. The satellite images, analyses and detailed ecological surveys provide an overview of the condition of the British countryside.

Diffuse pollution

Pollution caused by the general use or manufacture of certain substances, which does not readily lend itself to control at the point of use. Examples are the release of solvents into the air through the use of some paints and adhesives, or the discharge of phosphates from washing powders into the sewerage system.

Earth Summit

United Nations Conference on Environment and Development held in Rio de Janeiro in June 1992.

EC Birds Directive (79/409/EEC)

This applies to birds, their eggs, nests and habitats. It provides for the protection, management and control of all species of naturally occurring wild birds in the European territory.

EC Directive

A European Community legal instrument, binding on all Member States but leaving the methods of implementation to national Governments, and which must, therefore, generally be transposed into national legislation.

Ecosystem

A community of interdependent organisms and the environment they inhabit, such as ponds and pond life.

Endemic species

A species of animal or plant confined to a particular region or island and having, so far as is known, originated there.

Environmental appraisal

The process of defining and examining options, and of weighing costs and benefits before a decision, particularly one having significant environmental costs and/or benefits, is taken.

Environmentally Sensitive Areas

An agri-environment scheme, run by agricultural departments designed to promote traditional farming practices to protect and enhance the environment. The scheme operates in 43 designated areas of the UK where farmers and other land managers can enter into 10 year agreements to manage their land in designated ways to maintain and restore particular landscapes and habitats.

Ex-situ conservation

Conservation of species away from their natural habitats, eg long term storage in germplasm banks for future needs.

Extensive agriculture

A term generally used to signify the use of low input (ie without large scale use of agrochemicals in arable farming or high stocking densities in livestock farming) low output crop and livestock husbandry systems.

Eutrophication

Enhanced primary production of coastal inland waters through nutrient enrichment (typically from sewage discharge or agrochemical run-off) resulting in an imbalance of the normal flora and fauna associated with the area.

Farm and Conservation Grant Scheme

A scheme run by the Ministry of Agriculture, Fisheries and Food to help farmers maintain efficient farming systems and meet the cost of combatting pollution and conserving the countryside and its wildlife.

Farm Woodland Premium Scheme
A scheme under EC Regulation 2080/92 run by agricultural departments, which offers farmers payments over 10 or 15 years for the planting of new woods on agricultural land.

Fauna
All animal life.

Flora
All plant life.

Gene
The basic physical unit of inheritance of animals and plants.

Gene pool
The total genetic information possessed by the reproductive members of a population of sexually reproducing organisms.

Genome
Strictly, the set of chromosomes found in each nucleus of a given species. More loosely used to describe general genetic variation between species.

Genotype
The genetic composition of an organism.

Geomorphology
The study of the evolution of land forms, or the arrangement and forms of the Earth's crust.

Germplasm
Hereditary material transmitted to offspring via the gametes (male or female reproductive cells).

Going for Green
A committee, under the chairmanship of Professor Graham Ashworth, whose purpose is to get the message of sustainable development through to the man and woman in the street, and to influence them to change their lifestyles.

Greenhouse effect
Processes by which certain gases in the atmosphere behave like glass in a greenhouse; glass allows solar radiation which heats the interior, but reduces the outward emission of heat radiation.

Habitat
A place in which a particular plant or animal lives. Often used in a wider sense, referring to major assemblages of plants and animals found together.

Habitat Scheme
An agri-environment scheme, run by agricultural departments, which offers payments to farmers over 10 or 20 years to create and maintain specified wildlife habitats on agricultural land.

Habitats and Species Directive (92/43/EEC)
This requires Member States to take measures to maintain or restore natural habitats and wild species at a favourable conservation status in the Community, giving effect to both site and species protection objectives.

Indicative Forestry Strategy
The Department of the Environment's Circular 29/92 encourages local planning authorities in England and Wales to prepare Indicative Forestry Strategies to guide the development of forestry in their area. The production of such strategies is voluntary and their status advisory. Indicative Forest Strategies originated in Scotland and are currently widely used there.

In-situ conservation
Plants and animals in the wild state, in contrast to ex-situ conservation in zoos, botanical gardens or other collections.

Integrated Pollution Control (IPC)
An approach to pollution control in the UK, which recognises the need to look at the environment as a whole, so that solutions to particular pollution problems take account of potential effects upon all environmental media.

Intensive agriculture
A term generally used to signify the use of high input, high output crop and livestock husbandry systems in order to produce the optimum possible economic return from the available land which involve high usage of fertilisers and agrochemicals, mechanisation and so on.

Invertebrates
Animals without a backbone.

Livestock Quotas
Limits on the numbers of ewes or suckler beef cows on which a farmer can claim support payments. Milk support is limited by an annual quota on the number of litres of milk a farmer can sell without being subject to penalties.

Local Agenda 21 Initiative
An initiative set out in the publication Agenda 21: A Guide for Local Authorities in the UK.

Marine Nature Reserve
An area of the sea, including the seabed extending below low water, notified under the Wildlife and Countryside Act 1981

as being of special nature conservation interest. The designation applies throughout Great Britain and sites are notified by the appropriate Secretary of State.

Metadata
Data about data. A second level of information about other data sources. Thus a dictionary of all those who hold information on different groups of species could be described as species metadata.

Monoculture
Cultivation of only a single crop on a given area of land.

Montane
Above the potential tree line (at about 700m).

Moorland scheme
An agri-environment scheme, run by agriculture departments, which offers farmers on moorland in hill areas, outside the ESAs, payments over five years for reducing their sheep stocking densities and managing the land in order to improve the condition of heather and other shrubby moorland.

Morphology
The physical form of an organism.

National Nature Reserve
A reserve declared under law and managed either by one of the statutory nature conservation agencies or by an approved body.

National Park
Designated in England and Wales under the National Parks and Access to the Countryside Act 1949 for the purpose of preserving and enhancing the natural beauty of areas specified by reason of their natural beauty and the opportunity they afford for open-air recreation.

Natural Areas
These biogeographic zones reflect the geological foundation, the natural systems and processes and the wildlife in different parts of England, and provide a framework for setting objectives for nature conservation.

Nature Conservation Topic Centre
One of several international topic centres set up by the European Environment Agency, in this case to provide information and advice on issues relating to nature conservation and biodiversity in the countries of the Agency (the European Union countries plus Iceland and Norway).

Nitrate Sensitive Areas (NSAs)
An agri-environment scheme run by MAFF which aims to safeguard the future viability of selected public drinking water sources by compensating farmers for voluntarily changing their farming practices in ways which significantly reduce nitrate leaching.

Nitrate Vulnerable Zone
An area of land designated according to the provisions of the EC Nitrate Directive (91/676/EEC). The land forms the catchment of a surface or groundwater body which has been identified as a "polluted water" either on the grounds that a drinking water source exceeding 50mg/litre under specified monitoring conditions, or else that a water body, usually coastal or estuarine, is subject to specified undesirable ecological disturbance as a result of nitrogen-enrichment caused by agriculture.

Nitrogen dioxide (NO2)
A gas which mainly arises from fuel combustion in vehicles, boilers and furnaces, which is toxic in high concentrations and contributes to ozone formation and acid rain.

Nitrogen oxides (NOx)
A range of compounds formed by the oxidation of atmospheric nitrogen. Some of these oxides contribute to acid rain and smog, and can affect the stratospheric ozone layer.

Organic Aid Scheme
An agri-environment scheme, run by agriculture departments, which offers payments over five years to farmers who convert to organic production.

Photochemical reaction
Chemical reaction where the energy is supplied by sunlight.

Point source pollution
Pollution from specific sources which can be controlled directly, such as power stations whose emissions to the air can be controlled by scrubbing the exhaust gases, or through the use of cleaner fuel.

Pollutant
A substance which is present in concentrations which cause harm or exceed an environmental quality standard.

Polluter Pays Principle
States that the cost of measures decided by authorities to ensure that the environment is in an acceptable state should be reflected in the cost of goods and services which cause pollution in production and/or consumption.

Precautionary principle
Defined in the 1990 White Paper in the following terms "where there are significant risks of damage to the environment, the Government will be prepared to take precautionary action to limit the use of potentially dangerous materials or the spread of potentially dangerous pollutants, even where scientific knowledge is not conclusive, if the balance of likely costs and benefits justified it".

Protozoa
The simplest form of animal life; most are single cells and can be seen only with a microscope. They live mainly in freshwater, on sea shores and in soil.

Ramsar Convention
Originally agreed in 1975 to stem the progressive encroachment on, and loss of, wetlands, it requires the designation of Wetlands of International Importance.

Red Data Book species
Catalogues published by the International Union for the Conservation of Nature (IUCN) or by National Authorities listing species which are rare or in danger of becoming extinct globally or nationally. Sometimes species are included for which the national authority hosts a large part of the world's population, and has an international responsibility to conserve them.

Semi-natural habitats
Habitats or communities that have been modified to a limited extent by man, but still consist of species naturally occurring in the area.

Set-aside
Normally arable land removed from agricultural production as a requirement for receiving agricultural support.

Special Area of Conservation
A site designated by the UK Government under EC Directive 92/43 on the conservation of natural habitats and of wild fauna and flora.

Special Protection Area
A site designated under Article 4 of EC Directive 79/409 on the conservation of wild birds. Together SACs and SPAs form a network of European sites known as Natura 2000.

Species action plan
A conservation plan for a species based upon knowledge of its ecological and other requirements, which identifies the actions needed to stabilise and improve its status.

SSSIs - Site of Special Scientific Interest
An area of land notified under the Wildlife and Countryside Act 1981 as being of special nature conservation interest. The SSSI designation applies throughout GB. Sites are notified by the appropriate country conservation agency.

Statement of forest principles
Agreed at the Earth Summit in 1992. Countries are encouraged to prepare national plans for sustainable forestry.

Sub-species
A group of interbreeding populations with different characteristics (physical and genetic) from other populations of the same species, frequently isolated geographically from other populations of the same species.

Sustainable development
Development that meets the needs of the present without compromising the ability of future generations to meet their own needs.

Systematics
The study and arrangement of living things into groups as closely as possible according to their evolutionary relationships.

Taxonomy
The science of describing, naming and classifying living things.

Tir Cymen
A voluntary whole farm scheme for countryside conservation in Wales, designed and operated by the Countryside Council for Wales. Farmers may be offered annual payments in return for the positive management for the benefit of wildlife, landscape archaeology and geology and for providing new opportunities for quiet enjoyment of the countryside.

Vascular plants
Plants that have a vascular system, ie contain vessels for conducting liquids. This includes all of the flowering plants, conifers, and ferns and their allies.

Vertebrates
Animals with a backbone; mammals, birds, reptiles, amphibians and fish.

Wildlife Enhancement Scheme
A scheme run by English Nature for SSSI owners and occupiers to encourage positive land management for wildlife or geology.

ABBREVIATIONS

ADAS
Agricultural Development and Advisory Service

ASSI
Area of Special Scientific Interest (Northern Ireland)

BRC
Biological Records Centre (Institute of Terrestrial Ecology)

BTCV
British Trust for Conservation Volunteers

BTO
British Trust for Ornithology

CAP
EC Common Agricultural Policy

CBC
Common Birds Census

CCBR
Co-ordinating Commission for Biological Recording

CEE
Council for Environmental Education

CIS
Countryside Information System

CCW
Countryside Council for Wales

CFP
EC Common Fisheries Policy

DFE
Department for Education

DOE
Department of the Environment

DOENI
Department of Environment (Northern Ireland)

DOT
Department of Transport

DTI
Department of Trade and Industry

EC
European Community

EEA
European Environment Agency

EN
English Nature

ESA
Environmentally Sensitive Area

EU
European Union

FC
Forestry Commission

FCO
Foreign and Commonwealth Office

FHE
Further and Higher Education

FWAG
Farming and Wildlife Advisory Group

GB
Great Britain: England, Scotland and Wales

GIS
Geographical Information System

GMO
Genetically Modified Organism

IMO
International Maritime Organisation

IPR
Intellectual Property Rights

IT
Information Technology

ITE
Institute of Terrestrial Ecology

IUCN
International Union for the Conservation of Nature

JNCC
Joint Nature Conservation Committee

LAA
Local Authority Associations

LBAP
Local Biodiversity Action Plans

LEA
Local Education Authorities

LGMB
Local Government Management Board

MAFF
Ministry of Agriculture, Fisheries and Food

MNR
Marine Nature Reserve

NERC
National Environmental Research Council

NGO
Non-Governmental Organisation

NRA
National Rivers Authority (to become part of the Environment Agency on 1 April 1996)

PCB
Polychlorinated biphenyls

PPG
Planning Policy Guidance Note (England and Wales)

RSPB
Royal Society for the Protection of Birds

SAC
Special Area of Conservation

SNH
Scottish Natural Heritage

SOAEFD
Scottish Office Agriculture, Environment and Fisheries Department

SPA
Special Protection Area

SQL
Structured Query Language

SSSI
Site of Special Scientific Interest (Britain)

STW
Sewage Treatment Works

SWQO
Statutory Water Quality Objective

UK
United Kingdom: England, Scotland, Wales and Northern Ireland

UKBD
United Kingdom Biodiversity Database

Printed in the United Kingdom for HMSO.
Dd.301771, C35, 12/95, 3400, 5673, 340021.